Earth, Life and the Universe

...exploring our cosmic ancestry

●

Keith Tritton

Curved air
publications ltd

Published in the United Kingdom by Curved Air Publications Limited

Curved Air Publications Limited
Venture House, Boldero Road
Bury St Edmunds, Suffolk
IP32 7BS, UK

First Impression

ISBN: 0-9540991-0-9

© Keith Tritton 2001

British Library Cataloguing-in-Publication Data.
A Catalogue for this book is available from the British Library.

Typeset and printed by Premier Printers Limited, Bury St Edmunds, Suffolk

For Lindsay

Introduction

Does life exist anywhere else in the Universe – that is, apart from on Earth? If it does, what is it like? Is it anything like us or are we simply alone? Until quite recently, attempts to answer these age-old questions were little more than pure speculation, and hard facts were difficult to establish. Nevertheless, within the past few years we have found new planets orbiting other stars, discovered tantalisingly lifelike signs in meteorites from Mars and learned much about how life can survive in unexpectedly extreme environments. Today astrobiology has become a serious science, with new discoveries being announced almost daily. Both the United States and the European Space Agencies are planning space missions that will certainly expand our understanding of the subject further and may perhaps even discover the first signs of living extraterrestrial organisms in the next few decades.

How should this search begin? Where should we be looking? For what signs should we be searching? This book explains the present strategy. Always bearing in mind that extraterrestrial life is likely to have evolved differently, the study of life on Earth can inform us about the kinds of environment and raw materials that living things need in order to thrive and evolve. Then the science of astronomy can tell us where we might look for similar environments elsewhere in the Universe. Therefore life, the Universe and the connections and interactions between the two are the recurring themes throughout this book, drawing together the two very different disciplines of biology and astronomy.

My own background is astronomy, and I make no apology for both beginning and ending my story in the stars. The first part of the book is about the Earth's place in the Universe as a home for life – the only such home of which any of us is so far aware. It tells in general terms of the formation of the Earth and the development of the particular physical environment that made life possible on the Earth.

The second part looks at life on Earth, including the processes that power it, and traces its development on the early Earth. Here, the fascinating world of microbiology holds the key. All the elaborate varieties of modern life forms on Earth descend from simple organisms with which we still share many complex life processes. The amazing resilience of micro-organisms thriving in apparently hostile environments has led to a reassessment of the extent to which an Earth-like environment is necessary for life.

The Earth's constantly changing continents, oceans and climate act as a persistent force for the evolution of life and major catastrophes are responsible for mass extinctions and great changes in the course of evolution. In turn, life has had a crucial effect on the Earth, for example totally changing the composition of its atmosphere. Chapters 6 and 7 are about the inextricable links between the evolution of the Earth and the life it supports.

The last part of the book considers the possible environments and locations outside the Earth in which life might have developed and the kinds of steps that are being taken to explore them. Finally, Chapter 10 is where I draw together some conclusions and predictions about the nature of extraterrestrial life as well as speculate whether life on Earth is unique in the Universe.

I have assumed no previous knowledge of science on behalf of the reader and all scientific terms and ideas are explained as they are introduced. A glossary of the scientific terms used is included as an appendix along with a chart of the geological periods and eras since the Cambrian.

The twentieth century was the century in which our understanding of the Universe changed out of all recognition. It was the century in which we discovered the true nature of the stars, the colossal scale of the Universe and the evidence for its origin. It was the century of Einstein's theories of relativity and the Hubble Space Telescope. The twenty-first century will bring equally phenomenal changes in our knowledge. As part of this new revolution I firmly believe that the true standing of life in the Universe will be at last revealed.

Acknowledgements

I would like to thank the many people who encouraged and helped me to write this book, and especially Lynne Harrison, Gareth Lloyd, Martin Davies, Peter Willsher and Lindsay Simpson, who all read and criticised its first draft. I would also like to thank Tony Stanger for his sterling work on the illustrations and Andrew Sinclair for his expert advice on the Milankovich cycles.

Contents

Part I: The Earth and the Universe

1 The cosmic cycle

Nobody could deny the importance of the Sun in our everyday lives. It appears as a powerful symbol in almost every culture, taking its place in myth, ritual and holiday brochures, the giver of light and life.

1.1 - *Much of the radiation from the Sun is essential to the existence of living things on the Earth, such as trees and birds.*

Most people, if asked to say what images and impressions the Sun conjured up in their minds, would probably volunteer something like the following: it's round, yellow, hot and very bright. They might also mention, perhaps remembering an earlier painful experience, that it's the cause of sunburn. They might go on to add that sunlight helps make plants grow, that the Sun is responsible for the cycles of night and day and the seasons and is the driving force behind the weather. In fact they would probably be able to come up with a long list of other ideas as well as these.

As an astronomer, I might express the above ideas in a slightly different way, but most of them are basically astronomical observations. For example, the Sun is a source of radiation, of which the most familiar kinds are light – visible radiation – and heat – infrared radiation. And much of this radiation is essential to the existence of living things on the Earth. For instance, heat from the Sun keeps the Earth at a comfortable temperature and light enables plants to absorb energy through the process of photosynthesis. However, some kinds of radiation from the Sun – the near-ultraviolet radiation that causes sunburn is a good example – are potentially harmful.

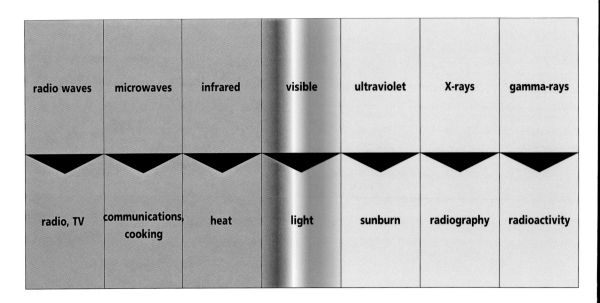

radio waves	microwaves	infrared	visible	ultraviolet	X-rays	gamma-rays
radio, TV	communications, cooking	heat	light	sunburn	radiography	radioactivity

1.2 - *The electromagnetic spectrum, with some of its uses and effects.*

Visible and infrared radiation – light and heat – are only a small part of a broad continuum that ranges from radio waves and microwaves through infrared, visible and ultraviolet radiation to X-rays and gamma-rays (Fig. 1.2). This continuum is called the electromagnetic spectrum, because its properties are governed by the laws of electricity and magnetism. All electromagnetic radiation travels through space with the same speed, the speed of light.

Although the Sun emits all of these kinds of radiation, it is most powerful in the infrared, visible and ultraviolet. As I have said, some parts of this spectrum help sustain life on Earth, but other parts can be quite deadly. Mid-ultraviolet radiation is lethal to many organisms because it breaks up the genetic material of the living cell and damages it beyond repair. Irradiation with ultraviolet light is an effective way of killing certain types of bacteria as well as a technique utilised to sterilise equipment used in food production and surgery. Solar radiation of this kind does not reach the surface of the Earth because it is absorbed in the atmosphere at an altitude of about 10 – 50 kilometres by the ozone layer. If the ozone layer were not in place, life as we know it on the surface of the Earth would be impossible. Ultraviolet radiation is one of the hazards of working in space, though the protective clothing worn by astronauts, who operate well above the ozone layer, blocks it easily.

The Sun is a star. That is, it is a huge, massive globe of extremely hot gas called plasma, almost completely composed of hydrogen and helium. The temperature at its surface is about 5,500°C (the temperature of an oxyacetylene flame torch is about 3,300°C) and it is an immensely powerful source of energy that is continually being radiated away into space. The rate at which energy is intercepted and absorbed by the Earth – about 140,000 million megawatts, equivalent to the output of 60 million large power stations – is of course only a tiny fraction of the Sun's total output.

Because we are accustomed to thinking of the stars as rather cold and remote it is easy to forget that the Sun is our nearest star. The Sun dominates our skies and our lives because it is so close, a mere 150 million kilometres away, very much less than a stone's throw in astronomical terms. The next star, known as Proxima Centauri, is a quarter of a million times farther away than this. But, like the Sun, every star is an enormous ball of high temperature plasma, pouring radiation out into space. It is only because of the vast distances that separate the stars from us that they seem faint and cold.

Since it is radiating so much energy away, does this mean that the Sun is cooling down? In principle it is a fairly easy calculation to estimate the total amount of heat energy contained in the Sun and so to work out how long it would take to radiate all this away. It turns out that this would take somewhere between 10 and 20 million years. Even more energy would be generated if the Sun were slowly shrinking under its own gravity, compressing and heating its own gases as it shrank, and it can be shown that this could extend the Sun's lifetime up to as much as 50 million years.

Towards the end of the nineteenth century, an estimate of 50 million years would have seemed quite consistent with the prevailing views of the age of the Earth. However, by the 1900s, this figure was starting to look decidedly problematic. Geologists were beginning to suspect that the Earth was far older than this, and the advent of radioactive dating methods eventually showed that its age was to be measured not in tens of millions, but in thousands of millions of years. The Sun could hardly be younger than the Earth. How was this paradox to be resolved?

Today, the best geological evidence tells us that the Sun's radiation cannot have altered very much over a period of at least 4,000 million years – we will see how this important conclusion was arrived at in Chapter 3 – and the true age of the Sun is about 4,600 million years. In fact astronomers have good reason to believe that, far from fading away, the Sun has actually grown a little brighter during this time.

Since the Sun has continued to radiate energy for thousands of millions of years even though its output has remained almost unchanged throughout this time, astronomers were forced to the conclusion that there had to be some process continually generating new energy inside the Sun. However, the source of this energy was not properly understood until the 1930s, when it was finally confirmed that the energy came from nuclear reactions taking place deep within the Sun's core.

Cosmic alchemy

In fact the Sun is an enormous nuclear reactor. As soon as this was understood, it became possible to make very detailed calculations about the interior of the Sun, for example by working out what the temperature and pressure must be at its centre. Its future evolution and expected lifetime could now be properly predicted. It was also possible to carry out the same sorts of calculations for the stars. However, before going down that particular avenue, I need first to divert slightly onto a subject that will turn out to be very closely related: the chemical elements. The chemical elements are very much a recurring theme throughout this book since these are a very important link between our twin themes of life and the Universe.

The chemical elements are the building blocks of matter from which all other materials are made. These elements combine with one another in chemical reactions to form compounds. Water, for example, is the chemical combination of the elements hydrogen (symbol H) and oxygen (symbol O) – two atoms of hydrogen with one atom of oxygen form one molecule of water (H_2O). When elements combine in this way, the characteristics of the compound are of course quite different from those of the constituent elements. Water, a liquid under everyday conditions, has quite different properties from either hydrogen or oxygen, which are gases.

The table gives a selection of the chemical elements. It is not complete in that about 112 have been identified altogether, but the table includes all of the first twenty, together with a few of the others whose names are familiar from a variety of different contexts and uses. The list is arranged in a well-established sequence that runs from the lightest to the heaviest elements. For each element, the table also gives a number called the atomic number, which identifies its position in the list, and its chemical symbol.

Atomic Number	Name	Chemical Symbol	Atomic Number	Name	Chemical Symbol
1	hydrogen	H	..	...	..
2	helium	He	26	iron	Fe
3	lithium	Li	27	cobalt	Co
4	beryllium	Be	28	nickel	Ni
5	boron	B	29	copper	Cu
6	carbon	C	..	...	..
7	nitrogen	N	47	silver	Ag
8	oxygen	O	..	...	..
9	fluorine	F	50	tin	Sn
10	neon	Ne	..	...	..
11	sodium	Na	78	platinum	Pt
12	magnesium	Mg	79	gold	Au
13	aluminium	Al	80	mercury	Hg
14	silicon	Si	..	...	..
15	phosphorous	P	82	lead	Pb
16	sulphur	S	..	...	..
17	chlorine	Cl	92	uranium	U
18	argon	Ar	..	...	..
19	potassium	K	94	plutonium	Pu
20	calcium	Ca	..	...	..

A selection of the chemical elements.

All terrestrial life works by making use of complex compounds of carbon called organic molecules. These compounds are almost entirely built out of only the lighter chemical elements: 99% of the atoms in the human body are either hydrogen, oxygen carbon or nitrogen, with hydrogen (60%) by far the commonest. Most of the hydrogen and oxygen is bound up in molecules of water. Apart from some very scarce trace elements, the human body consists of only these four elements together with much smaller amounts of sodium, phosphorous, sulphur and calcium and even smaller amounts of chlorine, potassium and magnesium. All of these are found amongst the lightest 20 elements in the table.

Hydrogen is the lightest of all the chemical elements. It is much lighter than air, which is mostly composed of nitrogen and oxygen, and so it was a natural choice for the gas that provided the lifting capacity of the early airships. However, hydrogen also has the property that it is highly flammable in air and it contributed to the notorious disaster that took place on May 6, 1937 in which 36 people lost their lives. The German transatlantic airship *Hindenburg*, 245 metres in length and the largest aircraft ever constructed, was coming in to land at Lakehurst Naval Air Station in New Jersey when fire broke out on board (Fig. 1.3). In seconds it was ablaze from end to end. Whether or not a gas leak was responsible for the fire was never established, but the accident more or less marked the end of the use of hydrogen in airships. Modern airships more commonly use helium, the next lightest element,

1.3 - *This image, taken at the time of the Hindenburg disaster, has been artificially coloured as part of a recent investigation into the causes of the accident.*

which although not as buoyant, is far safer than hydrogen because it is inert and incombustible.

Some of the chemical elements, like the metals gold, silver, copper, iron, lead, and mercury, have been known since antiquity, though most were isolated and identified only in the nineteenth and early twentieth centuries. The ancient alchemists experimented with transmutation, that is, the conversion of one material into another and in this sense at least they were the forerunners of modern chemists. The most famous of their goals was the transmutation of the so-called base metals into gold. However, we know today that gold is a basic chemical element, not a compound, and so it would never have been possible for the alchemists to achieve their aim by means of any chemical reaction.

Transmutation is, however, a feature of nuclear reactions. Nuclear reactions typically involve something like ten million times more energy than any chemical reaction, and so were plainly far beyond the capabilities of the alchemists.

There are two distinct types of nuclear reaction, the fission reactions and the fusion reactions. In fission reactions, the nuclei of heavier chemical elements like uranium or plutonium are split apart into smaller fragments. The resulting products are therefore different, lighter, chemical elements. The whole process is accompanied by the release of a large amount of energy. It was this kind of reaction that was used in the first atomic bombs, which are more accurately called nuclear fission weapons. A critical amount of fissile material, for example uranium or plutonium, is rapidly brought together, causing an explosion. As an illustration of the power of these weapons, the first nuclear fission bomb, the one dropped on Hiroshima in 1945, released energy equivalent to 15,000 tonnes (15 kilotonnes) of chemical explosive using as little as 60 kilograms of uranium.

All nuclear reactors presently used for power generation make use of fission reactions. A nuclear reactor is a device in which fission takes place under controlled conditions so that energy can be released controllably, not explosively. The fissile material, the fuel for these reactors, is again uranium or plutonium.

There is however a second, quite different, type of nuclear reaction; the fusion reaction. In fusion reactions, the fuel is a lighter chemical element like hydrogen or helium and the nuclei of these elements are fused together. This process therefore results in the build-up of heavier chemical elements. Again, large amounts of energy are released.

This is the kind of nuclear reaction that has been used in hydrogen bombs, otherwise known as fusion or thermonuclear weapons. In a nuclear fusion weapon, hydrogen fuel is ignited at a temperature of over ten million degrees Celsius generated by the prior explosion of a fission bomb. The actual fuel used is deuterium or tritium, a type of hydrogen sometimes referred to as heavy hydrogen. The yields of fusion weapons are much greater than fission weapons and can be as high as the equivalent of 20 million tonnes (20 megatonnes) of chemical explosive.

Thermonuclear fusion as a means of peaceful energy generation would have many advantages over the existing fission reactors. For a start, it would be a practically inexhaustible source of energy since the deuterium fuel is abundant and easily extracted from seawater. A fusion reactor would also be intrinsically much safer and less harmful to the environment than a fission reactor because it cannot undergo a runaway accident and it does not generate dangerous radioactive waste. As an additional benefit, nuclear reactors (of either type) make no direct contribution to global warming since they do not burn fossil fuels.

However, there are great practical difficulties to be overcome in building a power-generating fusion reactor, which requires the controlled confinement of a hydrogen plasma at temperatures of about 100 million degrees Celsius. An international collaboration of scientists and engineers from Europe, Japan, Russia and Canada is presently developing the International Thermonuclear Experimental Reactor (ITER), which is planned to be the first fusion device to produce thermal energy at the level of a commercial power station. In ITER, superconducting magnets confine and control the plasma and induce an electrical current through it. Fusion reactions take place only when the plasma is hot enough, dense enough and contained for a sufficiently long time. However, the technical problems are such that a commercial fusion power station is still some decades away from being realised.

To summarise, nuclear fission reactions occur when the nuclei of heavy chemical elements such as uranium or plutonium are split apart, whereas nuclear fusion reactions occur when the nuclei of light chemical elements such as hydrogen are forced together. Either process is accompanied by a large release of energy. Nuclear fusion converts the original fuel into a heavier chemical material: for example, hydrogen can be turned into helium.

The Sun and stars

The Sun is, in effect, a massive nuclear fusion power station. The main reason for this is the vast size of the Sun. Its diameter is 1,400,000 kilometres, over 100 times the diameter of the Earth, and its mass is 330,000 times that of the Earth. This great mass causes the gas pressure, density and temperature to increase enormously towards the centre of the Sun. The temperature at the surface is about 5,500°C but the temperature in the core is well over ten million degrees Celsius, and this is high enough for nuclear fusion to take place. The Sun is almost entirely composed of hydrogen (73% of the mass) and helium (25%) and it contains just a small percentage (2%) of other chemical elements, so the fuel – hydrogen – is available in abundance. The energy release, although large, is steady like a reactor, not explosive like a bomb, because the fusing material is confined to the core, the central 3% of the Sun, by the huge pressure of the overlaying gases.

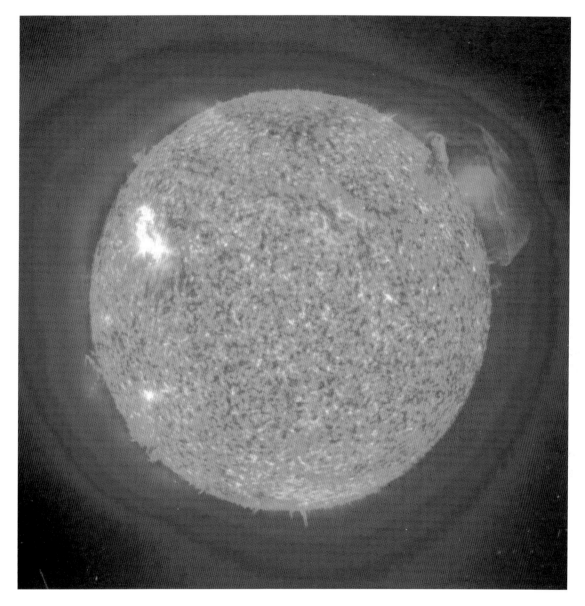

The particular nuclear reaction taking place is one in which hydrogen is being converted into helium at the astonishing rate of 600 million tonnes a second. Although this is enormous, it is possible to calculate that at the outset there was enough fuel in the core for nuclear fusion to continue for about 10,000 million years. This, then, explains the continued existence and constancy of the Sun over the last 4,600 million years – and also rather reassuringly shows that it will last for about another 5,000 million years to come.

1.4 - Activity on the Sun as seen from the solar satellite SOHO. The prominence at upper right is a huge cloud of relatively cool, dense plasma suspended in the Sun's hot, thin outer atmosphere. The hottest areas of the surface appear white and the cooler areas appear red.

We can now understand the long-term stability of the Sun. It is stable because the energy being radiated into space is replaced by the nuclear energy being generated in its core. The nuclear reactions sustain the high temperatures at the core, and these maintain the gas pressure that supports the weight of the overlaying material. Hence, the whole Sun stays in balance, neither shrinking nor expanding, remaining at the same temperature and brightness over an enormously long period of time.

In fact, to be strictly accurate, not exactly in balance, because as I mentioned above, it is thought that the Sun has slowly become slightly larger and brighter since its formation 4,600 million years ago.

The Sun, then, is a typical star. Astronomers have identified several types of stars, but the great majority of them, including the Sun, fall into a class called the main-sequence stars. These have many features in common with the Sun. They are all large, hot plasma spheres composed almost entirely of hydrogen and helium gas, with very small amounts of other chemical elements present, and they are all fuelled by nuclear fusion reactions consuming the hydrogen in their cores. However, individual main-sequence stars vary a great deal in size, temperature and especially brightness. The more massive stars are larger, brighter and hotter, and the less massive stars are smaller, fainter and cooler. The Sun is a fairly average main-sequence star, and we can use it as a convenient standard by which to compare the properties of other stars. This has been done in the table.

	Diameter (relative to the Sun)	Brightness (relative to the Sun)	Surface temperature (°C)
most massive	10 times larger	1,000,000 times brighter	50,000
Sun	1	1	5,500
least massive	10 times smaller	1000 times fainter	2,500

Properties of some main sequence stars.

It is now understood that all stars originally formed from large clouds of hydrogen and helium gas, pulled together and compressed by their own gravity. The gas slowly contracts in a process that takes perhaps 10 million years. The compression causes the gas to heat up, so as the cloud collapses, its temperature rises. Eventually the temperature

1.5 - The Eagle nebula, a region of recent star formation. At the centre, a young cluster of stars about two million years old illuminates the cloud of hydrogen gas from which they formed.

near the centre rises to several millions of degrees, creating just the right conditions for nuclear fusion to begin. The hydrogen ignites and a new-born star starts to shine.

What I have just described is the birth of a main-sequence star. Powerful astronomical telescopes show that star formation is still taking place all around us in stellar nurseries like the Eagle nebula (Fig. 1.5). As soon as nuclear fusion starts up, the new star stops contracting and it settles into a long period of stability, which in the case of a star like the Sun lasts 10,000 million years. But the more massive stars are hotter than the Sun and so the nuclear reactions proceed at a faster pace, consuming their hydrogen fuel at a furious rate. Because of this, their lifetimes are shorter – the most massive stars burn out in only a few million years. Conversely, less massive stars use up their fuel at a far more modest rate and most will outlast the Sun, perhaps surviving for 200,000 million years or more.

The hydrogen reserves in a main-sequence star are enormous, but they are not infinite, and finally the fuel in the core will be exhausted. This event marks the end of the star's steady main-sequence existence. What happens next is perhaps a little surprising. As the energy generation slows, the gas pressure in the core can no longer support the weight of the overlaying material. As a result, the star (or at least, its core) then starts to contract again, just as it had done at the very start of its life, and this contraction actually causes the central temperatures to rise even higher. In stars like the Sun, temperatures become high enough for a completely new fusion reaction to begin. Helium, which up until now has been a nuclear fusion product, now becomes the fuel in a new reaction, and the result is a new product: helium is converted into carbon.

The star now gets a new lease of life in which helium rather than hydrogen is the main fuel. However, this phase does not last as long as the main-sequence hydrogen-consuming phase because helium conversion does not release as much energy as hydrogen conversion. Inevitably the helium fuel in the core also eventually runs out. In the more massive stars the process of contraction and heating repeats itself and now carbon in turn becomes the next fuel, converting into oxygen. As the process continues, heavier and heavier elements are created; neon, magnesium and even iron.

This picture of the processes that are continually at work in the Sun and stars leads us to a remarkable conclusion. The stars are not only nuclear reactors, they are also chemical factories. The heavier elements are being manufactured in the cores of the stars, fashioned out of the raw materials hydrogen and helium. All the chemical elements from carbon to iron are created in this way, by nuclear fusion in the deep interiors of stars.

The dying stars

The end of a star's main-sequence lifetime is marked by the depletion of the hydrogen in its core. From now on, the star is doomed. In the dying stages of its life that follow, the star undergoes quite dramatic changes in its structure and appearance, as successive nuclear fuels are exhausted and new ones come into play. It is not easy to explain exactly why, but the shrinking of the core with each successive phase is accompanied by a huge expansion in the rest of the star. The outer layers of stars in these later stages of evolution swell larger and larger, becoming giant or supergiant stars. In about 5,000 million years time, our Sun will itself become a red giant star, expanding to something like 200 times its present size, engulfing the planets Mercury and Venus in the process. If the Earth survives at all, its surface will be blisteringly hot, with all life gone and the bloated red Sun filling most of the sky.

Supergiant stars are the final stages of the lives of the more massive stars, and these can swell larger still. The supergiant Betelgeuse, a prominent orange star in the constellation of Orion, is 1,000 times larger than the Sun. The outer layers of the red giant and supergiant stars become so hugely distended that they literally start to leak away into space.

1.6 - *The Helix Nebula, about 400 light-years away, is the nearest planetary nebula. It is the final stage in the life of a star similar to the Sun. The star's outer envelope has been blown away into space, leaving only the stellar core, which can be seen in the centre of the image.*

Ultimately, all the useful fuel in the core is used up and the star's life is over. Less massive stars (like our Sun) now shed their outer layers altogether in a mass expulsion of material called a planetary nebula. The material is dramatically thrown off in spectacular shells of glowing gas (Fig. 1.6). A substantial fraction of the material of the star is simply blown away into space. Only the tiny core of the star remains – in the figure it can be seen right at the centre – slowly cooling and fading away. Incidentally, these dazzling objects were given the rather misleading name of planetary nebula because many of them appeared spherical like a planet when viewed with a telescope. In reality they have nothing to do with planets at all, though the name has stuck.

The more massive stars end their lives even more spectacularly than this, in cataclysmic explosions called supernovae that blow the star completely apart (Fig. 1.7). The intensity of these detonations is so great that even heavier chemical elements – that is, elements heavier than iron – are formed in them, at the very moment of the star's destruction.

1.7 - *More massive stars end their lives in catastrophic explosions called supernovae that blow the star completely apart. This image shows the supernova remnant called the Crab Nebula.*

Whether in a planetary nebula expulsion or a supernova explosion, all stars end their lives by ejecting large masses of material out into space. This material consists very largely of gas from the outer layers and so it is mostly the unprocessed hydrogen and helium that has never taken part in nuclear fusion – the nuclear reactions take place deep in the hotter interiors of the stars and there is very little mixing between the core and the outer layers. However, the material does contain a small proportion of the processed material, that is, the heavier chemical elements. This material disperses widely between the stars and within perhaps about 100,000 years, a relatively short time in astronomical terms, it starts to mingle and mix with the existing interstellar gas clouds.

These are the very interstellar clouds that are themselves the material out of which new stars form. The process of collapse, heating and new star formation can now start all over again. Hence, material is continually being exchanged between the stars and the interstellar clouds as part of a cyclical process in which clouds collapse to form stars and stars eject material that rejuvenate the clouds. The very first clouds can have been composed of only hydrogen and helium, because these were the only elements in existence in the remote past, before the whole process of star formation started up. Therefore today, after innumerable loops through the cycle, the proportion of other elements has slowly crept up as the chemical factories did their work.

The most significant and surprising fact that emerges from this description of the cosmic cycle is that all the heavier chemical elements are created as part of this process, either in stars or in supernova explosions. This is the stuff from which all of the familiar material that surrounds us in our everyday lives is made. The remarkable conclusion is that, with the sole exception of hydrogen, the key chemical elements essential to life, the very elements of our own bodies, were created deep in the interiors of the stars as part of the cyclical process of stellar formation, evolution and rebirth. We are all made from stardust.

2 The emerging Earth

The primeval gas clouds from which the very first stars were formed contained no heavy chemical elements, only hydrogen and helium. How was this hydrogen and helium created? For the answer to this question we have to go back in time to long before even the Sun or the Earth existed, in fact as far back as the origin of the Universe itself.

The Universe is thought to have evolved into its present form out of an intensely hot, dense fireball called the Big Bang. It has cooled and expanded from this state over a period that must have lasted for about 13,000 – 15,000 million years. Perhaps not surprisingly, it has proved difficult to pin down the age of the Universe more exactly than this, and the question of its origin and evolution remains the subject of active research. However, despite this uncertainty, it has still been possible to work out quite a lot about the detailed circumstances of the Big Bang.

For example, it can be shown that all the primeval hydrogen and helium were created in the first few minutes of the Big Bang. It seems paradoxical that anyone can make such a definite statement about what happened in such a short instant of time during an event that occurred many thousands of millions of years ago, but this confidence comes from the fact that the prevailing temperatures at that time were so extremely high. In fact when the Universe was about one minute old, it is believed to have been at about 3,000 million degrees Celsius. In this intense heat, matter can exist only in a relatively simple state, and so understanding its properties becomes that much easier. As a matter of fact, at that time none of the chemical elements had yet come into existence because atomic nuclei were unable to form; the unimaginably high temperatures would have immediately broken them apart.

Although the conditions of the Big Bang cannot be reproduced in Earth-bound experiments, it is still possible to carry out a number of laboratory measurements that enable us to understand the behaviour of matter under such extreme temperature conditions. This is the province of the particle physicists, the scientists who study the fundamental subatomic constituents of matter, the elementary particles. It is rather paradoxical that the results of their research into the very smallest units of matter help astronomers to determine the history and fate of the entire Universe.

To carry out their research, particle physicists use high-energy accelerators, devices that produce and analyse beams of fast-moving atomic or subatomic particles. We are used to thinking of machines like these as atom-smashers, generating higher and higher energies and temperatures to break atomic nuclei apart, but in the Big Bang the whole process was the other way round – the nuclei formed out of their component parts, the protons and neutrons, as the temperature dropped. By the time the Universe had aged from one minute to ten minutes, it had cooled to about 500 million degrees. At this temperature, the protons and neutrons were able to combine to form atomic nuclei. All the neutrons fused with protons to form the nuclei of helium, and the protons that remained free and uncombined became the nuclei of hydrogen. Using only laboratory-based measurements it is possible to make a surprisingly accurate calculation of the ratio in which hydrogen and helium would have been produced in the Big Bang. It leads to the prediction that helium would have made up close to 24% of the mass. This is convincingly confirmed by direct observation of certain interstellar gas clouds today, where the proportion is found to be between 23% and 25%. The verification of this prediction is one of the greatest successes of the Big Bang theory and one of the main reasons for our confidence in it.

In passing, it ought to be mentioned that a minute amount of lithium was also created in the first few minutes of the Big Bang – though none of the heavier elements – and that the theory also correctly accounts for the amount observed today. However, their contribution to the following story is not significant, so I will refer simply to the hydrogen and helium gas clouds. It is not known exactly how these clouds condensed into star-forming regions, but it seems that stars began to form within about 1,000 million years after the Big Bang, and with the onset of star formation, the whole cosmic cycle started up.

In Chapter 1 I talked about the role of the stars as chemical factories, manufacturing all the heavier elements and, at the ends of their lives, recycling them back into the interstellar clouds. This means that the hydrogen is gradually being depleted and the heavier elements are slowly building up in the clouds. However, despite the truly massive rate at which the conversion of hydrogen to heavier elements is taking place inside the stars, and despite the enormously long time this has been going on, the entire amount converted is still only a very small percentage of the total. The interstellar gas clouds, the Sun and the other stars, indeed most of the Universe as a whole is still made up of roughly three-quarters hydrogen and a quarter helium, not very different from the way it was just after the Big Bang.

More exactly, the chemical composition of the Sun is as follows. Hydrogen constitutes a little over 73% of the mass, helium about 25% and carbon, nitrogen, oxygen and neon together constitute a mere 1.3%. The mass of all the other elements amounts to less than half a percent, and so only about 2% of the matter has so far been converted into heavier elements. These figures actually relate to the observable surface and outer layers of the Sun, material that has not taken part in the nuclear reactions in its core. So these proportions must represent the composition of the cloud of gas from which the Sun formed, and which already contained heavy elements that had been injected into it by previous stars. These proportions are also typical of the wider cosmos; almost everywhere we look in the Universe we find much the same distribution, with only small variations in the amounts of the heavier elements.

However, this simply doesn't accord with our everyday experience. The familiar world about us, here on the Earth, doesn't seem to be anything like this. The atmosphere we breathe is mostly nitrogen and oxygen. Ordinary building materials such as cement, bricks, and glass, like much of the rocky surface of the Earth, are made of minerals largely composed of silicates – compounds of silicon and oxygen. Obviously there must be a lot of hydrogen locked up as H_2O in the oceans, lakes and icecaps, and maybe even living things, but common sense tells us that this isn't likely to amount to anything like as much as three-quarters of the mass of the planet. Moreover, there is no obvious sign of helium at all. The proportions we see around us on Earth simply don't conform to the cosmic average.

At first sight it seems that despite the apparent uniformity of the chemical composition of almost all the Universe represented by the stars and interstellar clouds, our own planet is far from typical. To reconcile these observations we need to understand how the Earth – and the other planets – were formed.

Formation of the planets

The Solar System is the name we give to the family of planets, their satellites and various other smaller bodies like the asteroids and comets that are centred on the Sun. The Sun is overwhelmingly the most massive body in the Solar System, containing 99.9% of the mass, and so its gravitational pull dominates the system and all the other bodies orbit around it. In turn, nearly all the planets have one or more smaller satellites orbiting them, in the same way as the Moon orbits the Earth.

All these bodies, with the exception of certain comets, orbit the Sun in or very close to a narrow plane (Fig. 2.1). The view shown here is oblique – the true orbits are very nearly circular. All the planets move round the Sun in the same direction and the Sun also rotates in this direction.

Five of the planets, namely Mercury, Venus, Mars, Jupiter and Saturn, have been known since ancient times and all can easily be seen from the Earth with the unaided eye. Looking through even a modest telescope,

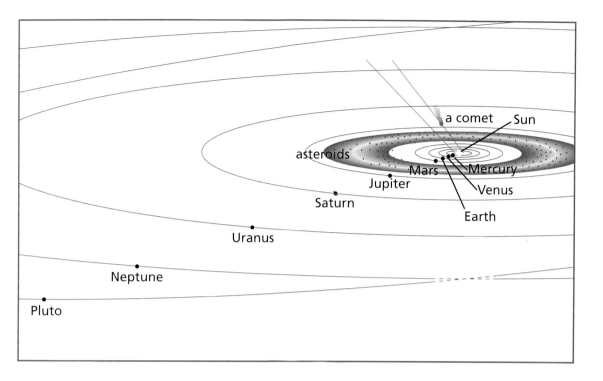

asteroids

a comet · Sun

Mars · Mercury

Jupiter · Venus

Saturn

Earth

Uranus

Neptune

Pluto

all of them will appear as a disc or crescent shape. Pale banded features can be seen running across Jupiter's surface, and its four largest satellites are clearly visible. So are the spectacular rings of Saturn, and its largest satellite Titan.

Galileo Galilei was the first person ever to use a telescope to observe the stars and planets. The series of discoveries he made in 1609 and 1610 were to transform the science of astronomy. He was able to resolve the enigmatic Milky Way into innumerable faint stars and he observed that the surface of the Moon was irregular, with mountains and craters. He showed that Venus went through a cycle of Moon-like phases and he was the first person to observe the satellites of Jupiter and realise that they were in orbit around the planet. These last two discoveries provided convincing evidence for the Copernican view that the Earth and planets orbit the Sun and helped establish the present model of the Solar System. Tragically, his defence of these ideas was to lead him into a lengthy conflict with the Catholic Church and permanent house arrest for the last eight years of his life.

2.1 - In this overview of the main components of the Solar System, the sizes of the orbits of the planets have been kept to scale, though the sizes of the Sun and planets have not.

The telescope eventually led to a rapid growth in the known numbers of members of the Solar System. Two more planets were added to the list; Uranus in 1781, and Neptune in 1846. In between these two came the discovery of Ceres in 1801, the first of the asteroids. The asteroids, also known as the minor planets, are a swarm of smaller rocky bodies whose orbits mostly lie between Mars and Jupiter. Ceres, a little over 900 kilometres in diameter, is the largest, but most are very much smaller, irregularly shaped and only a few kilometres across. More than 7,000 asteroids have now been observed and over 5,000 have well-determined orbits.

For four hundred years, ground-based telescopes continued to reveal more and more about the planets, but the greatest acceleration of our knowledge of the Solar System has taken place in the last 40 years, the result of direct exploration by space missions. Spacecraft have landed on, orbited or flown past the Moon, all nine major planets except Pluto, comet Halley, many of the satellites and a handful of asteroids. Images of their surfaces have been transmitted back to Earth and in some cases highly detailed maps have been compiled.

We now know that the major planets fall into two main groups. Mercury, Venus, Earth and Mars, the smaller planets of the inner Solar System, have rocky surfaces and are broadly similar in composition. With diameters ranging from 4,900 kilometres (Mercury) to 12,800 kilometres (Earth), these are collectively known as the terrestrial planets. On the other hand, Jupiter, Saturn, Uranus and Neptune are massive gas giant planets inhabiting the outer Solar System. Ranging in diameter from 49,500 kilometres (Neptune) to 143,000 kilometres (Jupiter), their solid surfaces are buried deep beneath vast gaseous atmospheres.

As well as these two major planetary groups, there is a host of small icy worlds. Many of the satellites orbiting the giant planets belong to this group, and Pluto is more like one of these than any of the other major planets. Is it possible to understand how this diversity arose?

The most widely accepted theory of planetary formation is derived from one that was put forward in its original form by Immanuel Kant in 1755. Alternative theories exist, and there isn't complete agreement on the exact details, but the theory does account for the main properties of the Solar System as we observe it. It is thought that the Earth, along with the other planets, formed at about the same time as the Sun, and from the same interstellar gas cloud. As the gas cloud from which the Solar System formed collapsed under its own gravity, the central region grew very dense and hot, eventually forming the Sun. The rest of the cloud flattened into a rotating disc. This cloud, called the solar nebula, would already have contained a sprinkling of heavier elements injected by previous generations of stars.

When nuclear energy generation started in the Sun, the collapse was halted and its temperature stabilised. Thereafter the disc of the solar nebula began to cool down slowly. At some point, material started to solidify out of the cooling gas in the form of small particles or grains. The kinds of solid particles that condensed out depended on the temperature of the region in which they formed, with minerals and metals condensing first. In the inner parts of the Solar System the temperature was always too high for water to condense as ice but in the outer Solar System icy substances – not only water ice but also ammonia and methane ice – did condense. These are known as the volatile materials.

2.3 - *The sixteen largest bodies of the Solar System to scale (excluding the Sun). Top row: the rocky bodies, including the four terrestrial planets, Io (a satellite of Jupiter) and the Moon. Middle row: the icy bodies. Ganymede, Callisto and Europa are satellites of Jupiter, Titan is a satellite of Saturn and Triton is a satellite of Neptune. At the bottom are the four gas giants.*

It might be expected that the composition of these grains would be similar to that of the solar nebula from which they formed, but only a small proportion of the hydrogen could have combined with other chemical elements to form compounds. Most of it remained in its gaseous state, and so did nearly all the helium, which hardly reacts with other chemical elements at all.

The solid grains clumped together into lumps of ever larger and larger units, slowly building up into substantial

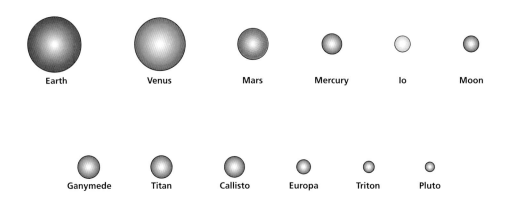

Earth   Venus   Mars   Mercury   Io   Moon

Ganymede   Titan   Callisto   Europa   Triton   Pluto

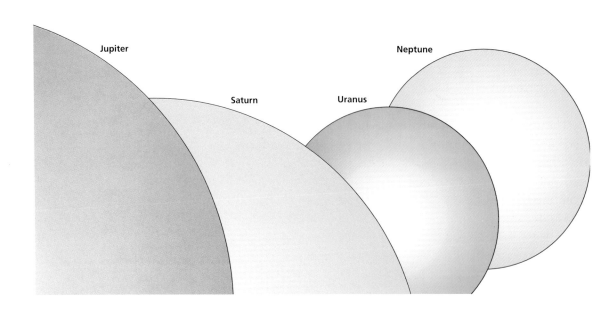

Jupiter

Neptune

Saturn   Uranus

mini-planets called planetesimals with enough gravitational attraction to attract other mini-planets. Eventually these accreted into planet-sized bodies in a process that, in the case of the terrestrial planets, probably took 100 million years or more. This period was one of collision, disintegration and re-formation with fewer and fewer but larger and larger bodies forming until the whole system settled down into much the state that we see it today. The debris from these collisions would have remained in orbit until it was swept up over time by the now established planets.

The terrestrial planets ended up with a preponderance of rocky and metallic materials as a result of the condensation process, since they were formed in the relatively higher temperatures of the inner Solar System. Bodies that formed farther out in the Solar System would in addition have accumulated a substantial quantity of the volatile, icy substances.

During the formation, another important process came into play. Each large collision would have generated enough heat to melt the embryo planet at the point of impact. In the case of a major collision, the planet might melt completely. The denser substances separated out of the melted material and sank to the centre of the planet. This is how the Earth, like the other terrestrial planets, developed a layered structure of a rocky mantle overlaying a core of iron and nickel. The icy bodies also developed differentiated layered structures, and by the same process, but in their case the layers consisted of ice overlaying a rocky core.

The residual solar nebula consisted mainly of hydrogen and helium, which did not condense. This eventually dispersed, but before it did so, a few of the larger bodies had become sufficiently massive to be able to attract gas from the solar nebula and retain it to form an atmosphere. These became the gas giant planets. The terrestrial planets were not massive enough to hold on to an atmosphere like this; hydrogen and helium, being light gases, can easily escape from their gravity. As we shall see, the origin of their atmospheres was quite different.

If this theory is correct, there must have been many small bodies orbiting the Sun during the early years of the Solar System and many collisions between them and the developing planets. In fact the impressions of these collisions can be seen on the Moon. The lunar craters first observed by Galileo and easily visible today with the aid of a pair of binoculars were caused by these ancient impacts (Fig. 2.4). Nearly all of them were produced on its surface during an intense phase of bombardment that ended about 3,800 million years ago.

2.4 - A cratered region of the Moon photographed from a manned Apollo spacecraft in lunar orbit. The large crater is called Aristarchus.

Evidence of impacts can be seen in the images of many other bodies too. The surface of Mercury looks much like the heavily cratered areas of the Moon. Mars bears many similar

scars, though it also has extensive plains where the cratering is quite light. Amongst the icy satellites, Callisto and Ganymede clearly show craters, though of course here they are formed in the surface ice, not in rock. One rather obvious exception to this pattern is the Earth. If the theory is correct, the Earth should have experienced the same bombardment during the early years of the Solar System, yet there are hardly any impact craters on the Earth. Why does the surface of the Earth look so different from these other bodies?

The Moon is a dead world, with no atmosphere or oceans, and it is now geologically inactive. Its surface has remained practically unchanged for thousands of millions of years. By contrast the Earth is a highly active planet, with a crust that has continually been formed and reformed by geological processes. The Earth would indeed have suffered the same bombardment as the Moon, but the scars of this phase would long ago have been obliterated by erosion from wind and water and also by geological processes. The interior of the Earth is still very hot, and this is the cause of continuing geological activity such as volcanism, continental drift and crustal reformation. There are other planets and satellites on which similar geological activity has partially or completely removed the bombardment evidence.

In summary, it is thought that the Earth, along with the other planets of the Solar System, formed as a by-product of the cosmic cycle at about the same time as the Sun, about 4,600 million years ago, and from the same material. However, our immediate environment on the surface of the Earth does not reflect the cosmic chemical composition of the Sun or the Universe as a whole, and this is for two main reasons. Firstly, the chemical selection that operated during the condensation process removed much of the hydrogen and nearly all the helium, and secondly the layering process resulted in materials of different chemical composition differentiating out into a core-mantle structure. In fact, provided we allow for the loss of the gaseous and volatile materials, the overall composition of the Earth does match that of the Sun reasonably well, and this lends support to the theory I have just outlined.

2.5 - The irregularly shaped asteroid Ida is about 52 kilometres in length. This image was obtained by the Galileo space mission from a distance of just over 10,000 kilometres.

Nearly all the condensed material of the solar nebula ended up in the major planets and satellites. However the asteroids (Fig. 2.5) remained small, never aggregating into anything larger. They were probably prevented from doing so by their neighbour, the giant planet Jupiter. Jupiter's huge mass constantly stirred up the orbits of the planetesimals orbiting between it and Mars, and much of the original material in this region has been scattered far and wide.

EARTH, LIFE AND THE UNIVERSE

Many smaller chunks of stony or metallic matter and lumps of condensed material continue to orbit the Sun. In fact the Earth continues to sweep up a staggering amount of this material, at the rate of some 200 tonnes per day. Much of it is burned up in the Earth's atmosphere, though a small proportion survives the fall to the surface of the Earth in the form of meteorites. An analysis of the meteorites that are recovered shows that many meteorites come from the asteroid belt, probably chipped off in collisions between them, though some of them come from the Moon and a very few from Mars.

2.6 - *Comet Hale-Bopp, one of the most spectacular to have appeared for many years, was visible in March 1997.*

Another important class of meteorite is known as the carbonaceous chondrites. These are chunks of primordial material that condensed directly out of the solar nebula but never became incorporated into a planet and so never underwent any heating or differentiation. It is found that their composition matches the Sun's very closely, apart from the elements that remained as gases and did not condense. This observation provides yet further confirmation of the theory and, as we shall see later, these meteorites also enable us to date the age of the Solar System itself.

A brief description of one more type of object, the comets, completes the inventory of the Solar System. Bright comets have appeared at irregular intervals in the skies throughout history. The periodic comet Halley, which was last visible in 1986, is perhaps the most famous, but the more recent apparitions of comet Hyakatuke in 1996 and comet Hale-Bopp (Fig. 2.6) in 1997 were more spectacular.

Comets are small bodies made of ice and dust, typically a few kilometres across, which were formed in the outer regions of the Solar System. They are remnants of the solar nebula that condensed a long way from the Sun and this is why they contain a high proportion of icy material; comets are made of some of the most pristine matter of the Solar System. They have highly elliptical orbits, occasionally darting through the Solar System's inner regions but spending the vast majority of their time in its remotest parts. It is only during their brief passage close to the Sun that a little of the ice and dust evaporates from their surfaces to produce the characteristic tails, tens of millions of kilometres long, that appear so spectacular from the Earth.

There may be as many as a million million comets in the Solar System and, despite their small size and the fleeting nature of their appearance, they play an important part in our story.

Life in the Universe

The Solar System is still only our backyard in astronomical terms. As I sit writing this in my office at home in a small English village, I have in front of me a battered old practice football which was abandoned by the children when they left home a few years ago. It measures about 20 centimetres in diameter. If we were to take this football to represent the Sun, then to the same scale we could imagine the Earth as an object about 2 millimetres across, perhaps a peppercorn, and it would be in orbit at about 20 metres from the football. The orbit might just fit in my garden. Pluto, which marks the outer limit of the planets, would be smaller, say a grain of sand, orbiting some 40 times farther, about 800 metres away, around the

2.7 - M51, the Whirlpool galaxy, a nearby spiral galaxy that contains something like a hundred thousand million stars. This is a composite of Hubble Space Telescope and ground-based images. It shows spiral arms and dust clouds as well as red patches of glowing hydrogen gas, which are star-forming regions like the one pictured in Fig. 1.5.

outskirts of the village. However, on the same scale, where would we find the nearest star, Proxima Centauri? In the next village? In the next county? No, on the same scale we would have to travel over 5,700 kilometres – it couldn't even be in Europe. To imagine Proxima Centauri to scale, think of another slightly smaller football, in the city of New York on the other side of the Atlantic.

However, this is just the nearest star. The Sun and all the stars we can see in the night sky with the naked eye are

2.8 - Part of the Hubble Deep Field, showing the faintest and farthest galaxies ever registered. The image was obtained by the Hubble Space Telescope over a period of ten consecutive days in December 1995.

part of the galaxy that we call the Milky Way. From a long way away, the Milky Way probably looks rather like the spiral galaxy pictured in Figure 2.7. Although it's impossible to see individual stars in this photograph, the galaxy is shining with the light of something like a hundred thousand million stars. It isn't really possible to grasp the significance of such a huge number, but it's about the number of grains of salt it would take to fill a small swimming pool. As it happens, a hundred thousand million is also about the number of galaxies in the observable Universe. Figure 2.8 shows an image of a minute piece of the sky obtained by the Hubble Space Telescope. If you made a hole in a piece of paper with a fine needle and then held the paper at arm's length, this is about the area of sky you would see through the hole. Almost every one of the hundreds of objects in the picture is a galaxy and each galaxy is shining with the light of thousands of millions of stars.

This picture of the Hubble Deep Field records the faintest and farthest objects ever registered, showing galaxy after galaxy full of stars. It is a kind of sample of the Universe, and there is good reason to believe that the Universe looks much the same as this in whatever direction we look. Since we see the same kinds of objects throughout the Universe, it seems that the same physical processes are going on and the same laws of nature are operating throughout the observable Universe. This is strong evidence that the cosmic cycle functions in the same way everywhere. So it seems very likely that the processes which led to the formation of the Sun and the Solar System have been repeated innumerable times throughout the Universe. In turn it seems likely that there are countless other stars with their own associated planetary systems. Indeed there may be countless Earth-like planets in the Milky Way, let alone the whole Universe.

We might go on to argue that unless the Earth is anomalous in some way, conditions suitable for the development of life might therefore exist on a multitude of as yet undiscovered planets. Since we know that life developed on one such planet, the Earth, what can we say about the likelihood that it did so on others, perhaps many others? One view might be that since there are a staggeringly large number of stars in the Universe, the statistical odds are heavily in favour of life having arisen elsewhere. However, this statistical line of reasoning begs a number of questions. How similar to the Earth must a planet be? Can we identify the crucial elements that made an environment favourable to the development of life? How certain can we be that life would develop and survive in such an environment? Moreover, can we say anything about how it might evolve? These are the questions I shall be trying to address in the rest of this book.

Since we presently know of only one place where we are sure that life developed, namely the Earth, it is impossible to be certain whether our planet is unique or one of many. We shall know the answer to this question for certain only if there does come a day when extraterrestrial life, perhaps thriving but perhaps extinct, is actually discovered. Although space probes have begun the exploration of the other planets of our Solar System, these have so far found no signs of life on them. It is true that very few space experiments have been designed with the detection of life as the main objective, though the environments that have been revealed on most other planets would appear to be rather hostile to the development of life. Mercury and the Moon are airless, the surface of Venus is at the temperature of melting lead, and so on. One series of experiments carried out by the Viking Mars missions in the late 1970s sampled the Martian soil in an attempt to detect biological material. The results were negative, but the experiments were rather limited and are not proof of the absence of life on Mars. More ambitious experiments are planned for the first decades of the twenty-first century and I shall talk about these in Chapter 8.

What about planets outside the Solar System? There are something like a hundred thousand million stars in the Milky Way. It is not known how many of these stars have their own planetary systems, but a significant fraction of them are similar in size and temperature to the Sun. The enormous distances between the stars (one football in

England, another in New York) means that it is impossible to pick up planetary systems orbiting other stars – extrasolar planets – simply by trying to see them through a telescope. They would be too close to their parent star and would be lost in the glare. However, other indirect techniques have recently established the existence of extrasolar planets around a handful of the closest stars. Only a few planets have been found, and it is too early to be very sure of the total numbers, but it could be that about a third of all Sun-like stars have their own planetary systems. Even so, there is as yet no direct evidence for the existence of Earth-like planets; all planets so far discovered are giants, more Jupiter-like than Earth-like. Moreover, we are very far from being able to say whether any of these support extraterrestrial life.

The questions we have asked ourselves are difficult to answer because we have scarcely begun to explore our own Solar System and we are as yet unable to make direct observations of any extrasolar systems. So far there is no evidence for the existence of life anywhere other than on the Earth. How then can we proceed? This is where the overlap of two different branches of science comes into play. Perhaps biology – terrestrial biology – can tell us about the environment and materials that life needs in order to thrive and evolve. And perhaps astronomy can tell us where to look for these environments elsewhere in the Universe.

Part II:
Earth and Life

3 The early history of the Earth

The formation of the Earth about 4,600 million years ago was the beginning of a period in which the planet suffered intense bombardment from the remaining planetesimals and the other debris that it was sweeping up from the early Solar System. During this phase, which lasted until about 3,800 million years ago, the crust of the Earth may well have melted and re-solidified many times. This immense timespan of about 800 million years, over a sixth of the entire age of the Earth, is the earliest period of geological time, known as the Hadean era. This was also the period when the Earth's metallic core separated out from its rocky mantle.

After this initial turbulent period most of the debris had been swept up and the impacts became much rarer, though they did not stop completely, and conditions on the planet became much more stable. The crust, or at least a part of it, solidified for the final time, forming the earliest terrestrial rocks. The oldest minerals known (zircon crystals) come from western Australia and are about 4,400 million years old, but the earliest rocks known, a mixture of sedimentary rocks and lava, have been found in the Isua hills in south-western Greenland and were laid down some 3,750 million years ago.

The ages of rocks like these are determined by radioactive dating. The phenomenon of radioactivity was first discovered and investigated by Antoine Henri Becquerel and Marie and Pierre Curie at the end of the nineteenth century. Certain compounds of uranium, radium and other substances were found to be giving out a low-level form of energy. Investigation showed that they were emitting positively and negatively charged particles and electromagnetic radiation in the form of gamma rays. The nuclei of the atoms of these radioactive substances were in fact spontaneously changing into other atoms (or other forms of the same atom called isotopes), ejecting these particles in the process.

This is a kind of nuclear activity, though it is not the same as the nuclear reactions I talked about in Chapter 1. It is a spontaneous process operating within the atomic nucleus and only certain atomic nuclei are radioactive. Some atoms are stable; those that can decay are called the unstable isotopes. These (parent) elements change or decay into other (daughter) elements at a strictly constant rate. In fact neither temperature nor pressure have any effect on the rate, and neither do electrical, magnetic or gravitational fields.

It is impossible to predict when any individual atom will decay. However, in any real sample there are a large number of atoms, and the overall rate at which they decay is measurable. The time taken for half the atoms in the sample to decay is known as the half-life of the substance. Imagine a bowl of apples and oranges in which apples change spontaneously into oranges in the same way as parent radioactive atoms decay into daughter products. At the start, assume the bowl contains only apples and no oranges. As time progresses, more and more oranges appear. After a certain time – the half-life – the bowl is half apples and half oranges. After twice this time the bowl is one-quarter apples and three-quarters oranges, and so on. The bowl becomes progressively full of oranges until eventually there are no apples left at all. When the rate of decay is known, the ratio of oranges to apples can be calculated at any time until all the apples have disappeared.

Nuclear physicists have determined the half-lives of a huge number of radioactive isotopes. Short-lived radioactive materials with half-lives of a fraction of a second can be created in laboratories. Nuclear explosions generate damaging radioactive materials such as caesium-137 and strontium-90 with relatively long half-lives, 27 and 28 years respectively. However, for many naturally occurring radioactive substances, half-lives can be measured in hundreds or thousands of millions of years; in fact by now they would have decayed away below the level of detectability if this were not the case.

In the 1900s Ernest Rutherford realised that the existence of these long-lived isotopes in the Earth's crust provided a reliable method of dating rocks. For example, uranium-235 decays to lead-207 with a half-life of 710 million years and uranium-238 decays to lead-206 with a half-life of 4,500 million years. In a mineral sample that contained uranium-235 but no lead-207 when the rock solidified, the age of the sample can be determined from the proportions of these two isotopes present today. (In practice, minor corrections need to be made for the small amounts of lead originally present.) As well as uranium-lead radiometric dating, other parent-daughter combinations are used such as rubidium-strontium and samarium-neodymium.

Although the Isua rocks, 3,750 million years old, are the earliest known terrestrial rocks, they are by no means the

3.1 - Three carbonaceous chondrite meteorites. On the left is the Allende meteorite that fell on Mexico in 1969. In the centre is a sample of a meteor that exploded over Yukon Territory, Canada, in 2000. On the right is a piece of the Murchison meteorite, which fell on Australia in 1969. Inset: Dr. Michael Zolensky, a NASA cosmic mineralogist, with a fragment of the Yukon meteorite.

oldest rocks on Earth. The meteorites called carbonaceous chondrites (Fig. 3.1) can also be dated using the same radiometric method. These meteorites crystallised directly from the primordial material of the solar nebula and have not re-melted since. They are therefore the oldest solid objects in the Solar System. All turn out to have similar ages of about 4,550 million years. This then is the age of the Solar System and therefore also the age of the Earth.

The Isua rocks provide us with a surprising amount of information about the early Earth. The presence of lava shows that the Earth was volcanically active at that time, and the pillow shapes of the lava formations are characteristic of underwater eruptions. In addition the presence of sedimentary rocks, which are formed from deposits of particles that have been eroded from the land by water or wind, shows that oceans and (probably small) continents were already established by the end of the Hadean era. The fact that large quantities of liquid water were present even tells us something about the surface temperature at that time; it was at least above freezing and below boiling point. This, incidentally, is the justification for saying that the Sun's radiation cannot have altered very much over a period of at least 4,000 million years; if the Sun had been significantly hotter or cooler then the oceans would have boiled away or frozen solid.

In some respects then, the landscape would appear familiar to us today. Volcanic islands and small, barren continents would be surrounded by oceans. But the Earth's atmosphere would have been quite different from today. It would have formed largely from the huge outpourings of gas that accompanied the volcanic eruptions of the Hadean era. The atmosphere would have consisted mainly of nitrogen, carbon dioxide (CO_2), water (H_2O) and sulphur dioxide (SO_2). The most notable difference from today's atmosphere is that, although oxygen formed a part of compounds like these, the atmosphere contained essentially no free oxygen at all. Now, ozone is in fact a different form of free oxygen – oxygen gas, O_2 is formed from two combined oxygen atoms and ozone is formed from three combined oxygen atoms, O_3 – and so there was no ozone in this early atmosphere either. The atmosphere lacked its protective filtering layer and the surface of the Earth was left exposed to a flood of solar ultraviolet radiation.

Despite this unpromising environment, the existence of both atmosphere and oceans was to prove crucial for the development of life. The Earth was massive enough to be able to retain this atmosphere and stop it from simply leaking away into space, and the presence of the atmosphere prevented the evaporation and loss of the oceans. The whole system achieved a long-term stability in which the surface temperature did not diverge outside relatively narrow limits right up until the present day.

Start of a long journey

There are three quite different kinds of geological evidence that record the signs of the very earliest known life. The first of these consists of the fossil structures known as stromatolites, found in the Barberton Greenstone Belt on the Swaziland-South Africa border and in the Pilbara region of Western Australia. The rocks from these regions were deposited about 3,500 million years ago.

3.2 - Living stromatolite reefs in Shark Bay, Western Australia. Inset: an underwater view.

Living stromatolites, though rare, can still be found today in a few rather remote locations such as Shark Bay in Western Australia. Stromatolites are colonies of bacteria with a characteristic mound-shaped structure up to about two metres across (Fig. 3.2). In cross-section they have a fine layered structure which is built up by the progressive deposition of sediments in shallow tidal waters. The bacteria in the top layer get their energy from sunlight through the process of photosynthesis. Lower layers are made of anaerobic bacteria; that is, bacteria that prefer an oxygen-free environment. As sediments are deposited, the bacteria migrate upwards and the sediments are left behind in layers.

The conditions in which stromatolites thrive, typically in intertidal zones, are ideal for fossilisation. Even after the bacterial colonies have died out, the structures remain as thinly layered mounds and the imprint of these layers is preserved as the sediment petrifies and forms an easily

3.3 - Photographs and drawings of 3,500 million-year-old fossil bacteria resembling modern cyanobacteria.

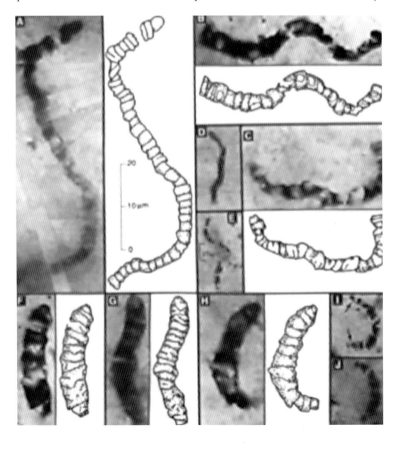

EARTH, LIFE AND THE UNIVERSE

recognised fossil. All fossil stromatolites, which occur in the geological record throughout the Earth's history, are much the same in size and structure.

The second kind of evidence for early life is provided by individual fossilised bacterial cells (Fig. 3.3). These microscopic fossils also date from about 3,500 million years ago. Certain of these, found within the Australian stromatolite fossils, have an appearance very similar in shape and size to modern bacteria called cyanobacteria. Cyanobacteria are amongst the largest of the bacteria, up to 50 microns long (one micron is 1/1000th of a millimetre). They are photosynthesising organisms containing pigments that absorb ultraviolet radiation. The Barberton Greenstone Belt rocks, which are about the same age, also contain fossil bacteria, rod-shaped objects just a few microns long, smaller than their Australian cyanobacteria cousins.

The presence within the stromatolites of cyanobacteria-like fossils suggests very strongly that similar photosynthesising bacteria existed 3,500 million years ago. They may even have helped the stromatolites to survive at a time when there was no ozone shield against the Sun's ultraviolet radiation. The stromatolites would have been partly protected by their semi-submerged existence in a shallow ocean since ultraviolet light is efficiently screened out by only a few centimetres of water. Moreover, the ultraviolet-absorbing cyanobacteria would have provided a blanket of protection for the other bacterial layers beneath them.

There is some evidence of life even earlier than this, though nothing so convincing as fossil remains. Carbon deposits of possible biological origin have been found in the oldest rocks of all, the 3,750 million year-old Isua rocks in Greenland. Analysis of the ratio of two naturally occurring isotopes of carbon, carbon-13 and carbon-12, hints at the chemical signature typical of living organisms. Although both isotopes participate in the same chemical reactions, the lighter carbon-12 is more easily extracted by certain living organisms, and they become slightly enriched in carbon-12. This enrichment is a feature of all fossils, including the stromatolites and cyanobacteria. Traces of carbon within the early Greenland rocks show an isotope ratio similar to that expected for living organisms. However, it is not completely out of the question that some non-biological chemical process could produce the same effect, and the evidence is not regarded as conclusive.

Nevertheless, it is clear that life in the form of bacteria of one kind or another was firmly established on the planet by about 3,500 million years ago and may even have been present 3,750 million years ago. The bacteria in the Barberton and Australian rocks are certainly the earliest life forms presently known, but it is highly unlikely that they are the earliest that actually existed or our earliest ancestors on the Earth. They are complicated organisms that must have already evolved from earlier life forms whose fossils have not been recovered and indeed may no longer exist. It is even possible that life established itself more than once during the Hadean era only to be obliterated each time by one of the large impacts.

Since the Earth's atmosphere contained almost no free oxygen at this time, the earliest life forms were all anaerobes, organisms that live in the absence of oxygen. They were all prokaryotes, which are microscopic, single-celled organisms containing no nucleus. It is now known that there are two types (known as domains) of prokaryote: the true bacteria and also a relatively recently recognised domain called the archaea (until a few years ago the archaea were classified with the bacteria, when they were known as archaebacteria). Both bacteria and archaea are very ancient life forms and both kinds survive very successfully today.

Bacteria thrive, often in extremely large colonies, in almost all natural environments, even including the intestinal tracts of humans and animals. In this sense, bacteria are perhaps the most successful and adaptable living creatures on Earth. Some bacteria cause disease, but most are harmless and some are even essential to higher life forms.

Archaea are usually found in naturally hot environments, such as hot springs, volcanic vents and boiling mud pools, but they also occur in human-made environments such as hot water tanks. Some cannot tolerate exposure to oxygen and some can only thrive at temperatures of 75 – 95°C or even higher. They grow on purely chemical nutrients in the absence of light and oxygen, often making use of sulphur.

Many bacteria also grow on inorganic chemicals but some feed on organic matter and others obtain their energy from photosynthesis, that is, the process by which an organism converts light energy into chemical energy. The cyanobacteria may well have been the first photosynthetic organisms. The development of photosynthesis in early Earth life forms was highly significant for two reasons. Firstly, photosynthesis provided a very efficient way in which the organisms could make use of an inexhaustible supply of energy, increasing their chances of survival. Secondly, and most importantly, oxygen is released as a by-product of the process, which uses sunlight to convert carbon dioxide and water to carbohydrates and oxygen gas.

However, in the beginning, hardly any free oxygen was retained in the atmosphere. Oxygen is a highly reactive gas that easily combines with minerals such as iron and sulphur in the oceans and the surface rocks and at first it was drawn out of the air as quickly as it was produced. Oxygen levels remained low for a very long time and therefore so also did ozone levels. Yet life did not only survive underwater or below ground. It colonised the land at least 2,600 million years ago. In the Eastern Transvaal, only a few kilometres from the Barberton Greenstone Belt, ancient soil surfaces of this age have been found covered with the remains of microbial mats that were probably originally several millimetres thick. It isn't known what kinds of organisms these early land-dwellers were, but they could have been photosynthesising organisms like cyanobacteria that manufactured their own ultraviolet-absorbing pigments.

Slow evolution

The process of evolution continued slowly, indeed very slowly. By about 2,000 million years ago the amount of oxygen in the atmosphere had started to creep up as the photosynthesising bacteria succeeded in producing oxygen at a faster rate than it could be absorbed by the oxygen-consuming reactions. Even so, the concentration had by this time still only reached about 1% of its present level. Nevertheless, it seems likely that the beginnings of an ozone shield would have started to form, and this would in turn have made the surface environment of the Earth a little more favourable to the advancement of life.

More complex organisms called eukaryotes (as distinct from prokaryotes) emerged at about this time. The earliest clear evidence for this is a spiral threadlike fossil of an organism called *Grypania*, which was found in an iron formation in Michigan, North America that dates from 2,100 million years ago. There are some even earlier signs of eukaryotic life; traces of complex cell membranes found in droplets of oil extracted from Pilbara shale in Australia could mean that they were in existence as far back as 2,700 million years ago.

Modern eukaryotes can be either single-celled, like the amoeba, or multi-celled, but even the single-celled ones are much larger than the prokaryotes. Unlike a prokaryote, each cell of a eukaryote contains a nucleus, holding the genetic information necessary for cell growth and reproduction. A eukaryote is a considerably more complex organism than a prokaryote, possessing tens of thousands rather than hundreds of genes.

After the bacteria and the archaea the eukaryotes form the third domain of life, called the eucarya. All multi-celled organisms belong to this domain, including the fungi, plants, animals and ultimately humans, though none of these branches evolved on Earth until very much later on.

All the first eukaryotes were small and single-celled. Their appearance was however associated with a new and huge evolutionary step forward. By now the concentration of free oxygen in the atmosphere, though still low, was sufficient to provide a totally new and much more efficient energy source – respiration. Respiration is the process by which organisms break down organic carbon compounds into carbon dioxide and water. Free oxygen is used up in the process and energy is liberated for the use of the organism. Only the cells of eukaryotic organisms were complex enough to be capable of this. The extra energy available resulted in a great evolutionary boost. Organisms grew larger and more diverse forms developed. For about 1,400 million years, single-celled eukaryotes were the dominant life form on the Earth.

Complex, multi-celled eukaryotes appeared about 600 – 550 million years ago, in a period called the late Precambrian. They developed specialised cells that were able to take on a range of form and function within the same organism. This inherent flexibility gave the eukaryotes a further evolutionary advantage, enabling them to diversify into a wide variety of different life forms.

By the start of the Cambrian period, which is taken to last from about 540 – 500 million years ago, the oxygen content of the atmosphere had built up to something like 10% of its present value. An ozone layer was in place and the scene was set for major changes in the evolution of the biosphere. An unprecedented expansion in the diversity of life took place beginning at the start of the Cambrian, an event sometimes known as the Cambrian explosion. Up until then, all life forms had been microscopic, but now the first shellfish, corals and animals with skeletons arrived in the seas. Many modern animal types appeared at this time, and it has been estimated that the number of kinds of animals doubled every 12 million years during this period. Not long after the Cambrian, fishes, vertebrates and land plants successively appeared until finally the land animals emerged about 400 million years ago.

The Earth's free oxygen level had by now stabilised to roughly its present value. A long-term balance was achieved between the photosynthesising organisms producing oxygen gas and the respiring organisms consuming it, with the atmosphere acting as the reservoir between them. Atmospheric oxygen is the product of biological activity, and in the absence of life it would slowly disappear as it was reabsorbed into the rocks and oceans of the Earth.

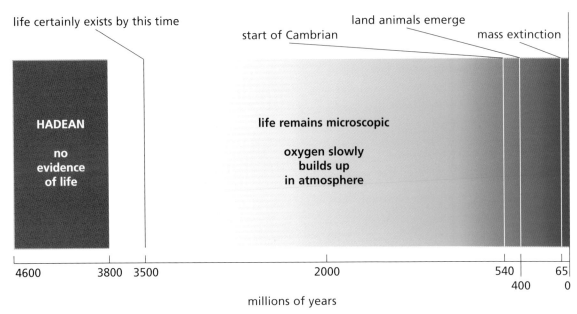

millions of years before present	
4600	formation of the Earth
3800	end of the Hadean era
3500	earliest unambiguous evidence of life
2000	oxygen levels reach 1% of present value
540	start of Cambrian explosion of life; oxygen reaches 10% of present level
400	first land animals; oxygen reaches present level
65	mass extinction of dinosaurs and many other species

3.4 - Key dates and events in the 4,600 million year history of life and the Earth.

Six steps

To help place these evolutionary milestones in context, I have drawn together some of the dates and events of the above story into a table and graph (Fig. 3.4) spanning the 4,600 million year history of the Earth. During the first 800 million years of this history, the period of intense bombardment known as the Hadean, we have no geological record and therefore no record of any life. We do not know when life first appeared on the Earth, but we do know that it already existed soon after the end of the Hadean and certainly by 3,500 million years ago. Whether life was able to retain a foothold even while the bombardment was in progress, whether it started up very rapidly as the bombardment subsided, or whether it developed and perished repeatedly during the Hadean, we do not know. What we do know is that at the time at which it does show up in the fossil record, it had already progressed some way down the evolutionary path, taking forms whose counterparts are still alive today.

There followed a vast period of time, about 3,000 million years, fully two-thirds of the Earth's history, in which life remained microscopic. Certainly, macroscopic structures existed, like the stromatolites, but these were formed by colonies of microscopic single-celled organisms, and were not themselves separate life forms. Large organisms with complex structure and specialised cells did not appear till about 600 million years ago. Life in the form, size and diversity that is familiar to us today developed only in this final phase, which has lasted less than a sixth of the Earth's history.

I am writing this chapter on the word-processor in my office. I am lucky that it is a reasonably spacious office; I have paced out its length, which is a little under five metres. I find I can pace from one end to the other in six good-sized steps. If the length of the office represents the whole history of the Earth, then my first stride takes in nearly all of the Hadean, the period of intense bombardment. Almost immediately after the end of my first step, life has already established itself on Earth but, although it diversifies and spreads wide, it remains microscopic and single-celled throughout the next four steps. Only during my sixth and final step does the living world even start to look like it does today, with fish in the seas and plants and animals on the land. The dinosaurs evolve but then abruptly disappear just over two inches away from the wall. On this scale, in which one millimetre represents a million years, where does the human race appear? It registers scarcely at all. *Homo sapiens* emerges only in the last tiny fraction of my final step, within a timespan represented by the thickness of the paper on the office wall.

4 Life on Earth

Nobody knows exactly when life first appeared on the planet, though it was well established soon after the end of the Hadean and certainly by 3,500 million years ago. Is it possible to say whether it has emerged anywhere else in the Universe? If it has, did its development run a similar or a different course there?

A standard observational approach in science, whether you are studying stars or bacteria, is to characterise the differences and similarities of a large number of examples of the object that interests you. For example, in Chapter 1 we saw that careful observation of a large number of main-sequence stars enabled astronomers to deduce that they came in a wide range of sizes and temperatures, but only a narrow range of chemical compositions. This result, together with many other parallel lines of investigation, eventually helped lead on to an understanding of the power sources, formation and life cycles of the stars.

But when it comes to the concept of 'life' in the sense in which I am talking here, we have only one example to work with. Of course, life on Earth is extraordinarily diverse, but it is nevertheless part of a single biosphere in which all the different species are interdependent and interrelated; how closely related we shall shortly see. In this sense it is unique, the only sample we know. Since all our knowledge is based only on this one sample, we need to be very cautious about coming to any conclusions about the nature of any other kinds of life there may be. It's as if we could observe only the Sun and no other stars; we would have a great deal of trouble deducing anything about the general properties or even the existence of other stars if this were the case.

To some extent the search for extraterrestrial life is an attempt to increase the sample size. If we ever came across any alien kind of life form, we would want to identify our differences and similarities and find out whether or not we had the same or a common origin.

There is a further important question which I haven't yet addressed. Although we have a general idea about the evolution of the variety of life on Earth from the earliest prokaryotes, there is still the problem of how life arrived here in the first place.

Readily acknowledging that alien life forms may be quite different, this is the point at which it is necessary to take a slightly more detailed look at the nature of life on Earth. However, I am not going to make an attempt at a precise definition of 'life'. This is partly because it's a difficult thing to do and partly because it has been better covered by other authors (see for example the works listed in *Further reading*) but mainly because it is not particularly helpful in advancing the theme of this book. Even if it were possible to agree on a form of words that perfectly described life as we understand it, it might not fit life on another planet. There is certainly no easy answer, no simple defining quality. Nevertheless, life possesses a number of attributes we do recognise. For example, living things eat, excrete and reproduce. They are complex yet organised. They grow, or at least develop. Although these are not uniquely defining qualities, they are properties upon which everyone would probably agree. It would be reasonable to summarise the most important of life's functions by saying that living things develop, order themselves and reproduce themselves. The name that is given to the complicated chain of chemical reactions that gather and use energy to carry out these functions is metabolism. Above all, life needs a source of energy to carry out these functions.

Even though life comes in every kind of shape and form and different kinds make use of quite different sources of energy, these are the main characteristics that are shared by all life forms. To what extent then do different life forms also share the same biological mechanisms for metabolism and reproduction? The answer is rather surprising.

Cells

For a start, all life forms are made up of one or more cells. The cell is the basic unit from which all living things are composed. Cells are complex entities. Even a single cell can be a complete organism in itself, as in all the prokaryotes and some of the eukaryotes. Indeed this was the only kind of life until 600 million years ago. However, larger organisms like animals can be made up of many millions of cells. The human body contains about 100 million million of them.

In the more complicated plants and animals, which are all multicellular organisms, groups of specialised cells become organised into tissues and organs. There are various cell types, for example skin, muscle, nerve and blood cells, but even so all cells in all organisms share a number of fundamental similarities of composition, form, and function.

A cell is a capsule of fluid called the cytoplasm, enclosed by a membrane that allows the cell to exchange certain substances with its surroundings. The cytoplasm is a watery liquid containing a mixture of large organic molecules that drive the cell and keep it working. The most important of these molecules are the proteins and the nucleic acids DNA and RNA. DNA (which stands for deoxyribonucleic acid) contains the genetic information that determines the essential character and hereditary properties of the organism; in fact DNA stores all the data needed to make an individual. DNA is also the means of transmitting the information from generation to generation. The function of RNA (ribonucleic acid) is to take this genetic blueprint and to manufacture the proteins, which are the real working chemicals of the cell. Proteins perform a wide range of the body's vital functions. For example, amongst other things proteins control the metabolism of the organism, act as chemical messengers between different parts of the body, transport material from one part of the body to another and protect the body from disease. Other proteins form structural components of the organism, like the protein collagen.

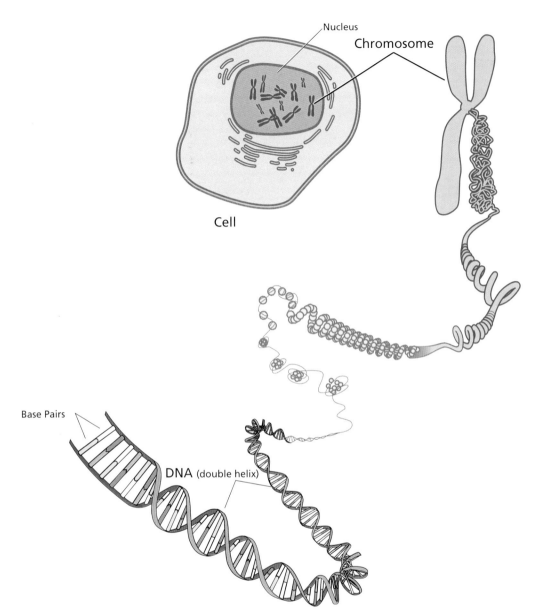

Nucleus

Chromosome

Cell

Base Pairs

DNA (double helix)

4.1 - *DNA is a very long thread-like molecule that is wound tightly round packets of protein and coiled tightly into dense bundles called chromosomes.*

DNA is a very long thread-like molecule wound tightly into a number of dense bundles called chromosomes (Fig. 4.1). If you could unravel the DNA from a single human cell it would stretch for about two metres, but it is actually packed into 23 pairs of chromosomes each only a hundredth of a millimetre across. Other species have different numbers of chromosomes. In any one organism, almost every cell preserves an identical set of chromosomes with identical DNA. In different species, the structure of the DNA is distinct but related; the more closely related the species, the more similar the DNA.

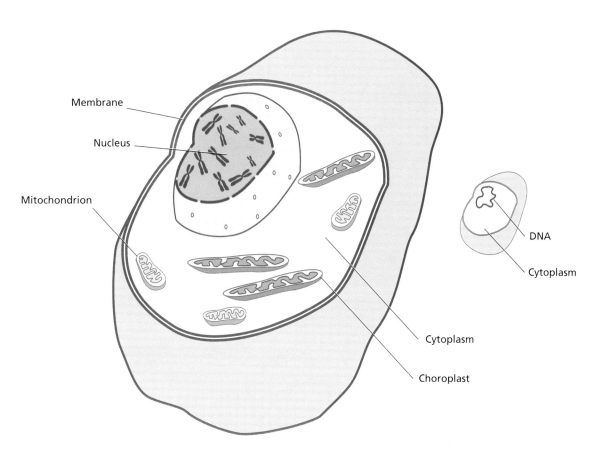

Membrane

Nucleus

Mitochondrion

DNA

Cytoplasm

Cytoplasm

Choroplast

4.2 - *Typical prokaryotic and eukaryotic cells. Prokaryotic cells are smaller and contain no nucleus. Eukaryotic cells contain energy-processing structures called mitochondria and (in plant cells, as here) photosynthesising units called chloroplasts.*

In eukaryotic cells, the chromosomes are contained within a membrane that forms a well-defined nucleus, the cell control centre (Fig. 4.2). Simpler prokaryotic cells like bacteria differ in that they have no nucleus and their genetic material is organised into a single chromosome. However, apart from these differences and despite the very different kinds of organism that they represent, all prokaryotic and all eukaryotic cells are very similar in composition and all function in very similar ways.

To get some feeling for how they function, it's probably best to start with the proteins, the workhorses of the living cell. To take just a single example, haemoglobin, one of the earliest proteins to have its structure determined, is a protein in the blood of many animals whose function is to transport oxygen to the tissues. The structure of haemoglobin is such that it is relatively easily able to combine with and detach itself from molecules of oxygen. It works by binding oxygen that it collects from the lungs and, after being transported elsewhere in the body by the circulation of the blood, releasing it where it is needed. When carrying oxygen, haemoglobin is bright red, and after releasing it, it is purple, giving rise to the different colours of blood in the arteries and veins. There are tens of thousands of different proteins in the human body, each with a specialised function of one kind or another.

All proteins are long, folded chains of amino acids. Amino acids are relatively simple molecules, compounds principally of carbon, hydrogen, nitrogen and oxygen with sometimes a trace of sulphur or iodine. More than 100 amino acids occur in nature, particularly in plants, but the vast majority of the proteins found in living organisms are composed of only 20 different kinds. These 20 amino acids are therefore some of the fundamental building blocks of life.

The exact structure of the amino acids need not concern us for the moment (see the glossary for more about this) but it is significant that they are made up almost entirely of carbon, hydrogen, nitrogen and oxygen, some of the simplest and most abundant chemical elements on Earth and in the Universe.

Proteins are quite complicated molecules, as would be expected, considering their role in the organism. They contain typically hundreds or thousands of amino acids strung together in a particular order. Each type of protein has its own unique sequence of amino acids, and this determines the shape and function of the protein.

Related species generally have similar proteins with similar functions, though the detailed structure of the proteins varies between them; for example monkey haemoglobin is slightly different from human haemoglobin. However, all these proteins are constructed from the same 20 amino acids; the variation arises only from differences in the number of amino acids and the order in which they occur.

These proteins need to be continually manufactured and renewed as the organism develops. They are manufactured by RNA according to the recipe specified by the DNA. RNA works by first copying a chunk of the DNA, which it carries out of the nucleus and into the main body of the cell, then interpreting the code and finally assembling the protein according to the instructions. To be exact, there are three main types of RNA; messenger RNA (mRNA), transfer RNA (tRNA) and ribosomal RNA (rRNA), each of which is responsible for one part of this process. One block of DNA instructions, specifying the complete assembly sequence of one protein, constitutes a gene.

The blueprint of life

DNA stores all the data needed to make an individual, but how precisely is this information stored? Both kinds of nucleic acid, DNA and RNA, are chains of molecules, and each link in the chain is called a nucleotide. The chain forms a backbone, and sticking out from the side of each nucleotide is a subunit called a base.

There are four kinds of nucleotide base in DNA, called cytosine, thymine, adenine and guanine, abbreviated to C, T, A and G. They can appear in any order along the length of the chain. As we shall see, the genetic information held by the DNA is encoded by the sequence in which they occur. RNA also has four bases, C, U, A and G, the same as DNA but with uracil (U) substituting for thymine. The nucleotide bases are compounds of typically 12 to 15 atoms of carbon, hydrogen, nitrogen and oxygen (see the glossary for more details) and so, like the amino acids, are made up of simple, abundant chemical elements.

The various types of RNA all consist of a single nucleotide chain, but DNA has a uniquely different configuration. It is actually a double chain with two strands coiled round each other into a double spiral or helix (Fig. 4.3). The two strands are linked together and spaced apart

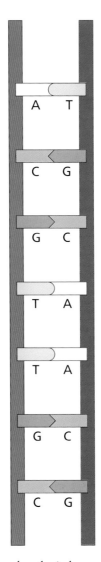

A	T
C	G
G	C
T	A
T	A
G	C
C	G

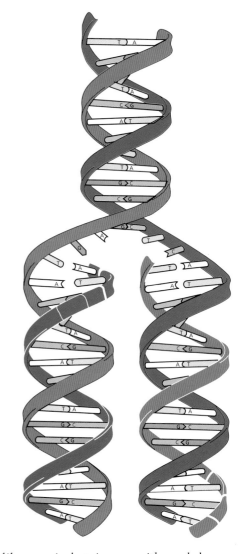

4.3 - On the left the DNA molecule is shown as a long chain of two strands coiled round each other into a double spiral or helix. The two strands are linked together and spaced apart by their bases. In the centre the coil is shown unwound. The bases are such that a C on one strand can only pair with a G on the other, and a T can only pair with an A. In replication (right) the helix unwinds and unzips, splitting into two separate strands. Each strand is now a complete template onto which a new complementary copy is built.

by their bases, rather like a spiral staircase with each base-pair forming one step. However, the shapes and sizes of the bases are such that a C on one strand can pair only with a G on the other, and a T can pair only with an A. The two strands are complementary, like a positive and a negative; if we know a sequence of bases on one strand, say A-C-G-T-G-G then we know the sequence on the other, T-G-C-A-C-C.

The molecular structure of DNA was worked out in 1953 by Francis Crick and James Watson working in Cambridge, and Maurice Wilkins and Rosalind Franklin (who died in 1958) working in London. For this work, considered to be one of the most important achievements of twentieth-century biology, Crick, Watson and Wilkins were jointly awarded the 1962 Nobel Prize for Physiology or Medicine. The realisation of this structure allowed the understanding both of how DNA worked and how it reproduced itself.

The translation from a sequence of bases on the DNA into a protein is accomplished because each adjacent triplet of bases codes for one amino acid in the protein. The bases are therefore read in groups of three. For example the

4.4 - The translation from a gene on the DNA into a protein. The DNA unwinds and unzips in the region of the gene to be activated. The negative strand acts as a template to which complementary nucleotides are attached, forming a length of messenger RNA (mRNA) that is a copy of the positive strand. The mRNA moves into the main body of the cell, where amino acids are placed into position in the correct sequence by transfer tRNA. The amino acids are finally clipped together into a chain by ribosomal rRNA, forming a new protein molecule.

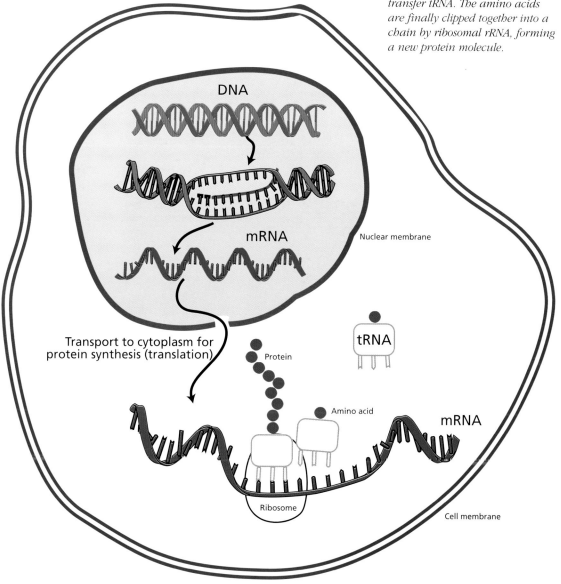

sequence C-A-U represents an amino acid called histidine and the sequence G-U-C represents one called valine. Since there are four bases to choose from, there are 4 x 4 x 4 = 64 possible triplets, or codons. There is even a 'stop' codon to indicate that the end of the sequence has been reached. Since there are only 20 amino acids, there is some redundancy in the coding; in fact most amino acids can be specified by more than one codon.

Mechanically speaking, how is the translation done? First, the DNA unwinds in the region of the gene to be activated, splitting the sequence of base-pairs to be copied rather like a zip fastener (Fig. 4.4). The unzipped negative strand acts as a template to which new, complementary nucleotides are attached, forming a length of messenger RNA that is an exact copy of the positive strand (except for the replacement of T by U). When complete, the mRNA moves into the main body of the cell. Here the appropriate amino acids are placed into position in the correct sequence by transfer tRNA following the code specified by each codon. The amino acids are finally clipped together into a chain by ribosomal rRNA, completing the manufacture of a new protein molecule.

Even the simplest DNA, say in a bacterium, contains a few million base-pairs and there are 3,100 million in human DNA. Genes coding for a single protein consist of blocks of several hundred adjacent nucleotides on the DNA; there are tens of thousands of genes in humans, typically only hundreds in a bacterium. But only about 3% of human DNA is taken up by genes; the remaining 97% is not used for encoding proteins. Yet its sequence is not entirely random and this part of the genome also seems to contain some kind of coded information. If so, the code and its function are as yet unknown.

The double structure of DNA is also the key to reproduction. When a cell divides into two, the helix unwinds and unzips completely, splitting into two separate strands. Each strand is now a complete template onto which a new complementary copy is built. The end result is two double helices, one destined for each cell.

Cell division continues repeatedly throughout the lifetime of an organism, generating new copies of the DNA and its genetic code at each step. If the copying process is perfect, as it nearly always is, then an identical copy is created. However, if only rarely, errors do creep in. These are called mutations. Perhaps one base in a gene is inexactly copied, leading to the wrong amino acid being used and an incorrect version of the protein being built. The protein may still work or it may fail to work. It may even work better, or at least better under certain circumstances. Mutations therefore result in variations, which can be advantageous but are more likely to be a cause of functional disorder. In the long history of evolution, mutations are one of the important sources of the variability through which natural selection acts, giving rise to the great diversity of life on Earth.

In June 2000, in a blaze of publicity and amid acclaim from heads of government and Nobel laureates, a consortium of 16 institutes from six countries constituting the Human Genome Project announced that they had accomplished the mapping of a working draft of 90% of the human genome, that is, the full sequence of 3,100 million base-pairs. The project had taken ten years to reach that point and the final draft was expected to take a further three. If ever printed out in full, the listing would be a string of Cs, Ts, As and Gs filling the equivalent of 200 telephone directories.

The significance of this achievement was that working out the complete genome sequence, or at least a substantial part of it, is expected to result in the identification of the sub-sequences that represent the individual human genes. In turn these should help establish the structure of the proteins these genes produce, leading to an understanding of how they work and how they might malfunction as a consequence of mutations. This knowledge could help determine an individual's predisposition to illness, guide the design of tailor-made drugs for treatment and allow the replacement or repair of faulty genes; gene therapy. Nearly 5,000 inherited disorders are known to be caused by mutations in a single human gene; for example cystic fibrosis and muscular dystrophy are single-gene defects and many others are at least partially induced by a combination of genetic factors.

Connecting to the tree of life

About 99.9% of the human genome is identical in all humans. Variations between individuals, not only visible differences such as the colour of hair and eyes, but also factors like susceptibility to certain kinds of disease, arise only through the fraction of a percent difference between their DNA. The study of these variations provides a wealth of information on human evolution and history, helping to map past human migrations and the relationships between different populations.

The DNA sequences of closely related species are very similar. In this respect human beings and chimpanzees differ by a mere 1.5%; it is this difference that makes us 'human' and chimpanzees 'chimpanzee-like'. The degree of difference is a useful measure of how closely related two species are. For example, human and mouse DNA differ by about 10%. This measurement can be made between quite distantly related species to find out, say, how different a human being is from a fruitfly or a nematode worm in genetic terms. The answer is about 40% and 60% respectively.

Taxonomy is the science of classifying and relating both living and extinct organisms. The traditional method has been to group organisms hierarchically according to their anatomical features, assuming that similarity of appearance or function is a consequence of a close evolutionary relationship. In this way we understand that for example humans are placed in a group with apes and monkeys, that these are all part of the wider grouping of primates, which in turn are classed with other mammals, then other animals and so on. This is interpreted to mean that humans share a relatively recent common ancestor with the apes but a progressively more remote common ancestor with other primates and other mammals.

The anatomical approach to classification was very successful but it also had its drawbacks, especially when it came to classifying micro-organisms. Comparisons of their rRNA sequences revealed that the single-celled bacteria and archaea are as different from each other as they are from the eucarya, and are not at all closely related despite their superficially similar appearance. In fact it was largely on the basis of the rRNA method, pioneered by the United States bacteriologist Carl Woese, that all life on Earth, living or extinct, was grouped into the three domains, bacteria, archaea and eucarya (or eukaryotes). The kingdoms that are more familiar to us, such as the plants, fungi, and animals, are subdivisions of the eucarya.

Genetic taxonomy is now replacing the more traditional anatomical taxonomy. Entire biological categories have been reclassified, often resulting in major rearrangements of previously accepted relationships between species. The underlying assumption is that the more closely two organisms are related, the fewer changes will have occurred to their rRNA sequences since they diverged from their common ancestor. Moreover, the degree of difference between them gives a measure of the time elapsed since this happened.

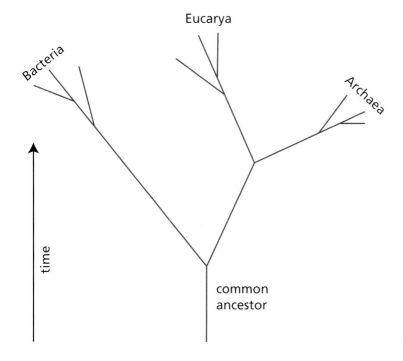

4.5 - *The three domains of life.*

This is not a precise dating method because the rate at which mutations occur is not expected to be constant. Nevertheless, the method confirms that the bacteria and the archaea diverged from a common precursor over 3,000 million years ago and that the eucarya and the archaea split into separate domains something like 2,700 million years ago (Fig. 4.5).

In summary, modern taxonomy bears out the hypothesis that every species is derived from another species. A useful analogue here is a tree with all its branches. The main trunk separates into various limbs, then divides repeatedly into boughs, branches and twigs. All individuals, living or not, can be placed somewhere on this tree and in principle we can trace the descent of any organism backwards to earlier and earlier ancestors by moving from the branches back towards the trunk. All present day life forms are cousins represented by the leaves at the ends of the twigs of the tree. Extinct life forms with no modern descendants are dead branches that don't extend through to the present.

Of course we can't analyse the genetic material of fossil organisms because it doesn't survive the fossilisation process, except in science fiction films like *Jurassic Park*. But we can still fit extinct species on to the tree of life by comparing them to their closest surviving relatives, or by using the anatomical classification system.

We draw the important conclusion that all life on Earth has a common origin. Biologists have catalogued something like 1,700,000 species, and it is certain that there are vastly many more yet to be discovered. Probably tens and possibly hundreds of millions of species exist on the planet. In terms of numbers of species, life on Earth consists almost entirely of insects and micro-organisms. For example, there are about a million known species of insect, though this figure could underestimate the total by as much as a factor of 10 or even 100 (J B S Haldane once famously remarked that God must be inordinately fond of beetles) and there are 40,000 known types of bacteria, though this is certainly only a very small fraction of the total. Even then, this is nowhere near the full tally since it is estimated that 99.9% of the species that have ever existed on Earth have already died out. And all these life forms, living or extinct, have a common ancestor.

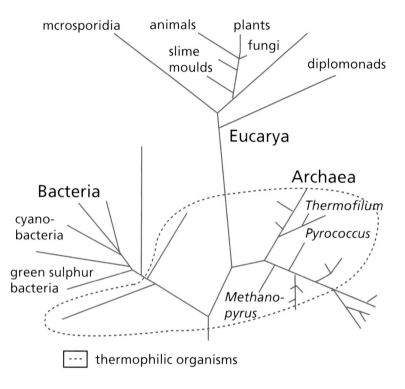

4.6 - *A representation of the tree of life, based on RNA sequencing. The least evolved organisms, near the base of the tree, are all thermophiles.*

Figure 4.6 shows a very small fraction of the tree of life, based on RNA sequencing. The length of each line represents the genetic distance that each kind of organism has travelled between branching and is a rough indicator of the time at which this happened. It shows the antiquity of the three domains of life and the relatively recent appearance of animals and plants.

In practice the true picture is more complicated than this. There has been genetic transfer between different branches of the tree; the Human Genome Project revealed that hundreds of human genes appear to have originated from bacteria. This means that evolution cannot be properly represented by a simple tree but is more like a web or net. Whether or not the majority of these transfers occurred before or after the split into the three principal domains is a matter of some controversy, though it is generally agreed that the mitochondria in eukaryotes and the chloroplasts in plants (Fig. 4.2) derived from bacteria that invaded early eukaryotic cells. The United States microbiologist Lynn Margulis believes that the eukaryotes may have originated from the fusion of early bacterial and archaeal cells.

The common ancestor

Incidentally, viruses do not have a place on this tree of life. They are far smaller and have a far simpler structure than bacteria and can reproduce only by injecting their genetic material into the cells of other organisms. They are a form of parasite in that no free-living kinds have ever been found. It is therefore a matter of definition whether viruses should be classified as life forms; most biologists do not include them on the grounds that they are not independent organisms. Either way, they do not fit on the same tree of life as everything else.

Leaving viruses apart, the nature of any life form is encoded by its genome, the particular sequence of DNA bases carried by its species. All organisms, living or extinct, from mammoths to mushrooms and bacteria to bananas, have utilised DNA and all of them have used the same genetic code. The same triplet of bases on the DNA was always translated into the same amino acid during protein manufacture. This code has not changed from the first known organisms to the present day.

However, there is a further vital connection that links everybody on the tree of life. It is called ATP.

We have already encountered three quite different ways in which organisms obtain the energy that they need to develop and reproduce. The archaea and some bacteria extract energy from inorganic chemicals, often based on compounds of sulphur. Green plants transform sunlight into internal energy through the process of photosynthesis. More complex organisms like animals use aerobic respiration, in which oxygen is used to break down organic carbon compounds into carbon dioxide and water. There are also other sources of energy, like fermentation.

In all these, the phosphorous compound ATP (adenosine triphosphate) is the common energy currency of every cell. ATP is formed when the cell acquires energy and it is broken down again at the point where the cell needs to use it up again. It is the mechanism that transfers biologically useful energy from energy-producing to energy-consuming reactions.

Despite the astounding diversity of life and the unimaginable range of time over which it has evolved, all the organisms on the tree of life share a remarkable number of characteristics. All are made of cells, all use ATP, all use DNA and all use the same genetic code. Even the earliest and simplest life forms known make use of the same complex chemistry as our own bodies.

We can trace our ancestry through DNA and demonstrate that all present day life forms are cousins to each other and that all life on Earth has a common origin. However, the link is even closer than that. We even share the most fundamental and sophisticated life processes with our common ancestor.

In fact we presumably have a series of common ancestors. Any life forms represented by the very base of the tree in Figure 4.6, before the first branching, are our common ancestors. Our *most recent* common ancestor is located immediately before the first branching but since this organism presumably evolved from something else, it would not be the first of them. It would therefore be more accurate to say we share the most fundamental and sophisticated life processes with the most recent common ancestor.

The genetic sequencing technique isn't able to identify any particular species as being the most recent common ancestor, because all known species have evolved away from it to a greater or lesser degree. However, it does show which species have evolved least from the most recent common ancestor. These species are all thermophilic, like many archaea, which means that their preferred environment is one in which the temperature is at or above 70°C. This tolerance of heat must be a characteristic of early life. Therefore the indications are that life either originated in, or since the most recent common ancestor is unlikely to be the earliest living thing, at least passed through a phase in a hot environment.

The origin of life

Did life on Earth originate as an oxygen-hating, sulphur-eating, single-celled creature in a hot spring, a volcanic vent, or deep underground? Or did it originate far earlier than we have so far been able to trace, perhaps in a totally different form, only to be forced through the bottleneck of a hot phase in the Earth's history?

It has to be admitted that nobody has yet been able to explain the origin of life. There is an enormous gap between the random jumble of raw chemical components in the early continents and oceans and the ordered complexity of an autonomous, self-replicating cell of proteins and nucleic acids, and nobody has a clear idea of how that gap could have been bridged by what was, presumably, a continuous process of development. Nevertheless, it is worthwhile considering some of the possibilities, not least because they reveal an unexpected astronomical connection.

All life that we are aware of is based on cellular units and one significant stage in its development must have been the formation of a cell wall. The wall is a membrane that separates the contents of cell and its surroundings, performing many essential roles. For example it keeps one organism's genetic material separate from that of another, it retains and concentrates the chemical materials necessary to keep its life processes going and it allows the inward absorption of necessary nutrients and the expulsion of waste. It must be selectively permeable but it must not dissolve in water.

Some possible mechanisms for cells to form naturally have been proposed. Certain organic molecules are known that can spontaneously form spherical membranes if heated in water. Amino acids can bond together into spheres that, although small, about a micron in diameter, are large enough to enclose other molecules in their interiors. Other more complex molecules called coacervates can also form larger hollow enclosures up to 500 microns across. However, modern cell walls are not made of either of these kinds of molecules and we would have to resort to the assumption that the original membranes were eventually replaced by the operation of some as yet unknown process.

It is even harder to imagine how the proteins and the nucleic acids might have developed. The main stumbling block here is their intimate interdependence. DNA is a passive medium that holds the blueprint for building the proteins but carries out no cell function except for replicating itself with the aid of proteins. The RNA can only read the blueprint and build the proteins that do all the hard work, but neither the RNA nor the proteins can reproduce themselves. And only the proteins can assemble the DNA and RNA in the first place!

The only way round this is to presume the existence of a simpler precursor system that evolved over time into what we see today. One school of thought has it that an earlier version of RNA alone could have previously done the job of both RNA and DNA. Since, like DNA, it is made up of a string of nucleotides, RNA could perhaps store genetic information on its own, acting as the genome without needing to read it from the DNA.

But even if RNA could exist without DNA, could it also exist independently of any proteins? In particular could it achieve its own replication? Experiments carried out in 1983 by two groups led by Thomas Cech in Colorado and

Sydney Altman at Yale University separately showed that isolated strands of RNA could become sufficiently active to cut and join other strands of RNA even without the help of proteins. It seems that the RNA predecessor might, albeit rather inefficiently, both carry its own genetic blueprint and reproduce itself unaided. Perhaps a concentration of this material could perpetuate itself, gradually evolving the separate, more efficient functioning of modern nucleic acids and proteins over a period of time.

There is an alternative view; perhaps the proteins came first. The problem here is that proteins can't reproduce themselves; or can they? It seems that certain protein fragments called prions may be able to do just that. Prions are the cause of a number of degenerative brain diseases in humans and other animals, including bovine spongiform encephalopathy (BSE) in cattle and Creutzfeldt-Jakob disease (CJD) in humans. A prion is an aberrant and highly resistant form of an otherwise harmless brain cell protein that can be acquired by infection or inherited from a mutation. It spreads by warping healthy proteins, inducing them to change shape and distort into the abnormal form. In this sense it reproduces itself, but it nevertheless remains a parasite and it is not suggested that such organisms could exist and survive independently. However, this behaviour is a further illustration of the capacity and flexibility of the protein form, and it may have played a role at some stage in the development of the mutual dependence of DNA, RNA and the proteins.

However this interdependence came about, and whichever of them came first, both proteins and RNA are rather long chains of molecules and they are not likely to assemble themselves spontaneously, certainly not in the absence of any biological activity. Is there any mechanism that might enable simple versions to form pre-biotically from what was presumably a dilute mixture of raw materials? Is there, for example, some naturally occurring template on which they could have formed? It has been suggested that certain mineral or clay surfaces with a microscopically regular configuration might attract and organise other molecules to form more complex structures.

Although this is a plausible idea, it remains a hypothesis for which there has as yet been no experimental verification.

It has to be admitted that our speculations about the origin of life are very sketchy indeed. The fact is that we are aware of only one way to create life, and that is by means of other living things. We assume the transition from inanimate to living things happened through a series of gradual and continuous processes, but we are very uncertain of what these were. However, it could be that the sequence of events was something like this.

Clay or crystal minerals would have been abundant on the early Earth as a result of the weathering of rocks, perhaps as sediment in the ocean or perhaps in the mud pools of a hot spring. Amino acids already present in the water or mud could have become attached to their regular surfaces and become assembled into molecular strings. Suppose that some of the resulting molecules had a limited self-replicating property, rather like crystals, and by chance some of them became encapsulated and concentrated within a membrane naturally formed from amino acids. Then the basic elements capable of eventually evolving into a living cell would now be in place.

Even though they were no more than an assortment of chemical blobs, some of them would grow and split faster than others because of their different internal chemistry. A kind of accidental competition would result, in which the most efficient replicators outpaced the others, at the same time passing on their chemical characteristics. The encapsulated molecules reproduced, changed and eventually specialised into RNA, proteins and eventually DNA, and in time the amino acids of the capsule walls were replaced by more effective cell membranes.

This story is only a suggestion, obviously incomplete and full of shortcomings. Moreover, there are alternative theories of how and where life might have begun, though nobody would yet claim to be certain of the true course of events. Nevertheless, if the process did actually get under way in some manner like this, it is not so difficult to imagine that it could have progressed, certainly undergoing numerous setbacks and running down many blind alleys, right up to the formation of the bacterial cell, at which point we are able to pick up the story in the fossil record.

Even this sketchy picture depends on there being a supply of molecules at least as complex as amino acids to hand. As it happens, this may not be too much of a problem. There is an ample supply of them, though they come from a rather surprising source.

Raw materials from space

You will recall that there is strong evidence that the Earth, along with the Sun and other planets, formed from an interstellar gas cloud. Radio astronomers have discovered the signals of quite complex molecules in these clouds, including familiar ones like water (H_2O), ammonia (NH_3), formaldehyde (HCHO) and vinegar (acetic acid, CH_3COOH). There's even an inexhaustible supply of alcohol (CH_3CH_2OH) in each cloud, countless millions of litres of it as a matter of fact, although it's rather inconveniently spread out in the form of a very thin gas. In order to get enough to fill a whisky glass you would have to condense it from a volume as big as a planet. Well over 100 different molecules like these have been found, produced by chemical reactions that take place naturally within the clouds themselves.

Significantly, the amino acid glycine has also been detected in these interstellar clouds. It forms from a chemical reaction between acetic acid and ammonia, and the clouds may well contain other amino acids yet to be discovered. Another molecule of biological importance found in the clouds is glycoaldehyde, a simple sugar molecule that is one of the subunits of the nucleic acids. It has even been suggested that molecules as complex as the nucleotide-base adenine may form naturally in this interstellar environment. In Chapter 1 I likened the stars to chemical factories. Now it seems that the interstellar clouds could be vast chemistry laboratories, assembling the basic building blocks of life from the raw elements created deep inside the stars.

How could these vital molecules could have reached the Earth? Could they have been swept up during the collapse of the cloud from which the Sun, Earth and other planets were formed? Not directly, because they are too volatile. The high temperatures of the inner solar nebula, the region from which the Sun and Earth formed, would have destroyed them. Nevertheless, they could have survived in the more remote outer regions, where the temperatures remained low. And, they could have been brought to the Earth at a later stage, when the Earth had cooled, by comets or meteorites – remember that the comets originate in the outer parts of the Solar System. It would be quite ironic if comets, historically seen as an omen of bad fortune, were actually part of the chain of circumstances through which life started on the Earth.

There is tangible evidence for this theory; amino acids have been discovered in meteorites. One Sunday morning in 1969, a large meteorite fell out of the sky near the town of Murchison in Australia, breaking up into many fragments as it did so. About half a tonne of material was recovered. It belonged to the class of carbonaceous chondrites, composed of primordial material that condensed directly out of the solar nebula. A variety of organic compounds were found to be present in the meteorite, including several of the protein-forming amino acids.

Now it happens that the biologically produced amino acids found in living things can be distinguished from those formed by non-biological processes, for example by synthesis in a laboratory or in the interstellar clouds. This is because amino acids exist in two forms, the L-form and the D-form. Since they are three-dimensional configurations of atoms or groups of atoms, they can exist in either of two arrangements that are mirror images of each other, rather like a pair of gloves. In effect, there is a left-handed and right-handed version of each amino acid. This property is called chirality.

Living things (on Earth) can use amino acids only in the L-form, whereas synthesised amino acids occur in both forms in equal proportions. Despite their chemical similarity, only the L-forms will work properly in biological processes. In theory, somewhere in the Universe there could be a fully functioning life system just as complex as the terrestrial one but based only on the D-forms. These two systems would be completely incompatible with each other. The chiral unity of all terrestrial life is yet another strong indicator of its common ancestry.

In the Murchison meteorite, both forms occur almost equally, though there is a slight excess of the L-form. This tells us a great deal. Firstly, we can rule out any suspicion that the samples are simply contamination by terrestrial biological material after the meteorite fell. Secondly, it confirms that amino acids can be created by natural (but presumably non-biological) extraterrestrial processes. Their presence is therefore strong support for the suggestion that the building blocks of life were manufactured in the interstellar clouds and were subsequently transported to the Earth in meteorites and comets.

In case you think that the fall of a meteorite is such a rare occurrence that it is unlikely to contribute significant amounts of organic material to the planet, bear in mind that an astounding 200 tonnes of cosmic debris is swept up by the Earth every day. Most of this is dust, only about 5% of which is in the form of carbonaceous chondrites, but even if these contain only 3% organic molecules, this represents an accumulation of some 100 tonnes per year. Moreover, this flux could have been as much as 100 times higher in the past.

As an alternative to this idea, might there be ways in which large organic molecules could have been manufactured naturally somewhere on the Earth? In the 1950s, Harold Urey and Stanley Miller carried out a famous series of experiments at the University of Chicago. They attempted to re-create what they believed to be the conditions in the primitive Earth atmosphere by enclosing a mixture of gases in a sealed flask and passing electric discharges through them to simulate lightning. After about a week, the flask was analysed and found to contain a rich blend of organic molecules that included some of the 20 protein-forming amino acids and also even the nucleotide bases adenine, guanine and thymine.

The difficulty with these experiments is that they were built around the assumption, current at the time, that the

4.7 - A black smoker, a hydrothermal vent on the ocean floor. The black cloud is formed by minerals precipitating out of a plume of superheated water as it mixes with the cold seawater.

Earth's early atmosphere was a mixture of methane, ammonia, hydrogen and water vapour. As we saw in Chapter 3, this is very different from what is now thought to have been the case. Similar experiments based on a more realistic gas mixture have not been very successful at producing amino acids and this process no longer seems convincing as a supply of the molecules required to assemble the first cells. The main significance of these experiments was to demonstrate that a natural non-biological way of producing organic molecules did exist, even if it wasn't in fact particularly effective in the environment of the primitive Earth.

Recently, a much more promising suggestion has been put forward. In 1979, during a deep diving submarine expedition in the middle of the Pacific, a joint French-US team discovered a previously unknown kind of volcanic activity on the ocean floor. It can be seen as a series of plumes of superheated water, heated deep underground by hot rock and magma, gushing out of the ocean floor. Where the hot water mixes with the cold seawater, minerals precipitate out, creating a muddy cloud known as a black smoker (Fig. 4.7). The crust leaks a steady supply of hydrogen, methane and hydrogen sulphide into the water. These are exactly the sorts of raw materials needed to produce organic molecules, and the porous nature of the seabed in the vicinity of the black smokers is ideal for concentrating them and promoting the right chemical reactions. In fact laboratory simulations of this environment show that it is far more efficient at producing organic molecules than the Miller-Urey method. I shall be saying much more about these black smokers, or hydrothermal vents, in the next chapter.

To summarise, it is still not certain how the organic molecules that were needed for the origin of life came to be present on the Earth. However, we have at least two plausible mechanisms; one a confirmed extraterrestrial source and one depending on ocean floor hydrothermal vents which must surely have existed in the Hadean. Probably both these, and maybe others too, played a part.

The raw materials, then, probably existed in abundance on the early Earth. And we understand, in general terms at least, how the modern diversity of life evolved from our common ancestor. Nevertheless, the level of complexity in even the simplest bacterium is far greater than a few amino acids strung together. Precisely by what route a collection of organic molecules organised themselves into the earliest living things and then into the DNA-based world of living cells is by no means clear. It is this intermediate stage of evolution that remains the greatest challenge to our understanding of the origin of life.

5   Life at the edge

In 1987 the United States Department of Energy was carrying out a programme of subterranean research motivated by the possibility of storing nuclear waste in very deep underground caverns. As part of this programme, investigators drilled a series of deep holes in the grounds of the DOE's own nuclear establishment at Savannah River in South Carolina. Very much to their surprise, they discovered a wealth of microbial organisms happily thriving in the sedimentary rocks beneath them down to a depth of 500 metres and more.

This was the first proof that anything could live at this sort of depth below ground. Until then, it had been widely believed that any subterranean micro-organisms would be confined to the topsoil, within one or two metres of the surface. Not only that, but the drills had brought up an enormous variety of microbes; thousands of different strains were retrieved, many completely new to science.

This discovery prompted searches in many other locations, each experiment probing greater and greater depths and revealing more about the previously unsuspected living world beneath our feet. Bacilli were recovered from a three-kilometre borehole in Virginia. Minute cracks within solid granite and basalt bedrock were found to harbour living micro-organisms. And even samples of sediment obtained from 750 metres beneath the ocean bed were found to be richly populated, with up to ten million bacteria per gram of material.

It seems quite possible that there are bacteria *everywhere* on Earth where the temperature is less than about 100°C. This represents a huge volume of the crust, and if it were true it is conceivable that the subsurface biomass, that is the mass of all the subterranean micro-organisms, is as great as all the living material on the surface, including all the animals and plants.

The deepest investigation so far conducted has been at the bottom of a South African gold mine, three and a half kilometres below ground. This was found to be the home of a thermophilic bacterium belonging to the genus *Thermus*, relatives of which are also found in a variety of warm environments on the surface. Moreover, this example was living in fissures in the solid rock, where the natural temperature is above 60°C and the pressure is 400 atmospheres.

No multicellular plants or animals nor any single-celled eukaryote can survive sustained temperatures as high as this, let alone the pressure. Yet it seems that many archaea and some bacteria not only survive, but positively thrive in these conditions. We've already come across these creatures; they are the thermophiles, the species that have evolved least from the most recent common ancestor.

The most extreme heat-lover yet discovered inhabits the tower-like vents of the black smokers on the mid-ocean bed. An organism known as *Pyrolobus fumarii* lives in the walls of these volcanic outcrops at temperatures that have been measured at a staggering 113°C.

These and similar discoveries raise a host of questions. If an organism can tolerate temperatures as high as this, to what other physical extremes can it adapt? Can life exist, for example, in regimes of very low temperature and pressure, or very high acidity? In these capabilities and many others, the bacteria and archaea outstrip everything else on the planet.

Some like it cold

Lake Vostok, one of the world's largest freshwater lakes, is half a kilometre deep, with a surface area of 14,000 square kilometres. However, it doesn't appear on many maps. It lies four kilometres under the Antarctic ice sheet directly below the coldest spot on Earth, Russia's Vostok research station, where the temperature fell to a record −89°C in 1983. Biologists who were keen to know whether it was possible for any kind of life to survive in the lake started to drill down and retrieve samples of the ice lying immediately above it. Great care was needed not to introduce any kind of contamination into the lake and so the drill was not allowed to break through into the water itself. By this method frozen lake-water samples were obtained from within 120 metres of the surface, some of the deepest ice cores ever recovered. Sure enough, even here there were signs of microbial life, a species of rod-like bacteria. The water in the lake has probably been cut off from both air and sunlight for 25 million years.

5.1 - Ice core samples at the Vostok research station in Antarctica.

Cold, darkness and isolation, then, do not seem to be insuperable barriers to life. Other discoveries confirm this; bacteria have been found at the bases of glaciers, living in the dark within thin layers of icy water generated by the friction of the glacial motion, and certain Antarctic bacteria can grow and divide at temperatures of −17°C or below.

Nor does the presence of air or indeed any kind of atmosphere seem to be necessary. In November 1969 the Apollo 12 crew set their lunar module down on the Moon only 200 metres away from the landing site of an earlier unmanned spacecraft, Surveyor 3, which had arrived in April 1967. The astronauts were able to retrieve certain equipment from the craft and bring it back to Earth. On examination, one of the cameras was found to have been accidentally contaminated by a colony of the bacterium *Streptococcus mitis*, which was still thriving after its round trip from Earth despite having been left on the Moon in a vacuum for well over two years. Another colony of the common soil bacterium *Bacillus subtilis* survived unprotected on an orbiting spaceship even longer, for nearly six years.

More bizarre still are the creatures that enjoy nothing better than bathing in concentrated acid. Some bacteria prefer vinegar and lemon juice, but the strangest of all was unearthed by a research project in Australia in the 1940s that was prompted by a serious problem with public sewer pipes. The pipes in question were made of concrete, yet as little as two years after they had been installed they suffered extensive corrosion and in some cases collapsed altogether. The problem was eventually traced to a micro-organism called *Thiobacillus thio-oxidans*. This bacterium lives on sulphur, and in these pipes it was extracting it from the hydrogen sulphide gas given off by the sewage. But *Thiobacillus thio-oxidans* contentedly produces, lives in and indeed requires concentrated sulphuric acid in which to grow. The acid strength is enough to dissolve metal and it was of course the acid that was destroying the concrete.

A close relative, *Thiobacillus ferro-oxidans* not only generates sulphuric acid but also tolerates high concentrations of iron, copper, cobalt, nickel and zinc. The ability of organisms like these to absorb poisonous metals and convert them to harmless compounds might one day be exploited by humans, for example to clear sites that have been contaminated by industrial waste. *Thermus*, the thermophilic inhabitant of deep mines, can also digest heavy metals and it has been speculated that bacteria might even have been responsible for concentrating and forming seams of mineral deposits.

In summary, life – especially microbial life – seems to have adapted to be able to survive in an extraordinarily wide range of ecological niches. Oxygen is essential for human life, but there are anaerobes that are indifferent to it and some that are actually poisoned by it. Some bacteria can resist temperatures of over 110°C and pressures of several hundred atmospheres or endure the cold and vacuum of interplanetary space. Some can live in extremely acidic and others in extremely alkaline environments.

Of course, no single organism can tolerate all these extremes. Even the sulphuric acid-loving *Thiobacillus* is killed off by an equivalent concentration of, for example, hydrochloric acid. Each organism, through the process of biological evolution, has undergone its own series of adaptations with the result that it suits the particular environment it now occupies. And all of them have evolved from our most recent common ancestor.

Returning for a moment to the thermophiles, we recall from the previous chapter that the more primitive known life forms are all thermophilic. From this we deduced that in evolutionary terms the thermophiles are close to the most recent common ancestor and therefore that this common ancestor probably inhabited a hot environment, meaning one having a temperature of 70°C or more.

This does not necessarily mean that life on Earth originated in this sort of environment, although that is certainly one possibility. The mud pools of volcanic hot springs have been put forward as the kind of place where life might have arisen, and the hydrothermal vents of the ocean floor have been suggested as another. Not only are these places at the appropriate temperatures but they also had the potential to supply the right organic raw materials. Moreover, it is very likely that these environments existed on the Earth soon after the end of the Hadean era. If life did develop in such a place, then it subsequently diversified and evolved to exploit the wider and generally cooler range of environmental niches that it occupies today.

An intriguing alternative possibility is whether the hot stage might have occurred even earlier, actually during the Hadean era itself. Enormous amounts of heat were generated by the giant impacts of planetesimals on the Earth during this period. The largest impacts would have melted the crust entirely, but as the intensity of the bombardment diminished, the crust would have solidified and the oceans would have formed. In fact the oceans may have formed and reformed many times in the intervals between the later, less violent collisions. Under these conditions it might be expected that deep-living organisms would survive better than any that preferred to live near the surface. Their refuge would be in the hot rocks a kilometre or more below ground. Harsh though these conditions seem, they are no stranger than those in which we find microbial life flourishing today.

Of course, the very earliest life forms need not have been thermophilic at all. It could be that an enforced passage through a hot phase meant that only the thermophiles survived. This opens up the question of where life originated, a question we shall postpone until Chapter 10.

What does life need?

Now let us look at this extraordinary adaptability of life from another point of view. We've seen that life is able to evolve to survive in a whole range of different environments. But, what is it that renders an environment habitable? What exactly does life need out of its surroundings? The answers to these questions could provide us with clues to the nature and possible location of extraterrestrial life and the circumstances of the origin of life on the Earth.

No living organism can be totally independent of its environment, at least not while its life processes continue to operate, though some microbes are able to enter a state of dormancy in which their vital functions are suspended for long periods of time. Metabolism, the series of chemical reactions that drives growth and development, requires energy. That energy has to come from the surroundings, and so does the carbon and the other raw chemical materials that are needed to manufacture the structures, fluids and complex molecules that make up the living cells and which need continual regeneration.

Humans eat food to acquire both energy and raw materials. Mostly we eat plants or other animals that eat plants. Plants, which convert solar energy to food by photosynthesis, are the primary food source. But, what does a bacterium feed on? To take an extreme example, what does a bacterium that lives three kilometres underground in solid rock feed on?

Living organisms make use of three basic sources of energy and matter: sunlight, inorganic chemical compounds and organic compounds. Those that acquire carbon by feeding on organic matter, such as sugars, proteins, fats, or amino acids, are called heterotrophs. Humans and other animals are heterotrophs, but many micro-organisms are also heterotrophic, ultimately depending on photosynthesis and sunlight for their existence.

Other organisms obtain their carbon from the inorganic compound carbon dioxide (CO_2). These are called the autotrophs. The life forms that use photosynthesis to generate cellular energy from light energy are one type of autotroph, collectively known as the phototrophs. Plants are the most obvious examples of phototrophs, but again many bacteria, like the cyanobacteria, are phototrophic.

Finally, there is a type of autotroph that has the ability to extract both matter and energy from purely inorganic compounds. These organisms can exist in the absence of both light and organic food, indefinitely isolated from the surface (the bacteria that live in rock belong to this group) obtaining their energy from reactions with inorganic salts. These are the chemoautotrophs. Certain deep-living methane-generating archaea get their energy from naturally occurring hydrogen and carbon dioxide gas, living out an existence that is truly independent of life at the surface or the products of photosynthesis. These methods of extracting energy are relatively inefficient, and so these organisms develop only slowly.

In some cases these organisms constitute the base of a food chain that supports entire biological communities. As we know, the mid-ocean hydrothermal vents harbour an abundance of bacterial life, but they are often surrounded

5.2 - A colony of giant tube worms near a hydrothermal vent.

EARTH, LIFE AND THE UNIVERSE

by substantial colonies of macroscopic life as well. Sometimes the surrounding area is teeming with life, including shrimps, fish and animals called tube worms (Fig. 5.2), which are centimetres across and grow up to a few metres in length. These are isolated communities, living in total darkness, feeding on each other and on the microscopic organisms whose primary source of nourishment is the minerals emerging from the vents.

In these vents, water is heated as it circulates beneath the sea bed, surging back into the oceans at temperatures of up to 380°C (at a depth of about two and a half kilometres, the extreme pressure prevents the water from boiling). It cools rapidly as it mixes with the water of the ocean floor, which is at about 2°C. Some of the dissolved minerals quickly precipitate out, generating the plumes of black smokers, and some are ideal as energy sources for microscopic life forms. Bacteria live in the walls of the volcanic outcrops and also within the plumes themselves, somehow managing to manoeuvre themselves into a comfortable zone in between the temperature extremes.

5.3 - *A spider from the Movile Cave.*

There are other examples of isolated ecosystems like this. In 1986, while excavating for a construction project in Romania, engineers accidentally broke into a cave that had no natural entrance and had until then been entirely cut off from the surface. This is Movile Cave, near the Black Sea coast. Its atmosphere is very rich in carbon dioxide and poor in oxygen. In its lower part there is a small lake, abundant in hydrogen sulphide, part of the groundwater table.

This cave houses an extraordinarily rich self-contained ecosystem which, in total darkness, is entirely sustained by inorganic chemical energy. It is inhabited by over 40 species of terrestrial and aquatic invertebrates including worms, snails, leeches, spiders and water scorpions. Over 30 species are new to science and live only in this cave (Fig. 5.3). It is thought that these creatures are the descendants of ancestors that were isolated some 5.5 million years ago, since when they have evolved and adapted to their underground existence in a number of ways that mark them out from their surface-dwelling cousins.

The animals are blind, having lost all sensation of sight in the persistent darkness. They have had to adapt to this sightless existence, moving around, feeding and recognising each other in the dark. All their surface pigmentation has faded, so they appear pallid and colourless. Moreover, they have adapted to survive on bacteria and fungi that in turn derive their energy from the sulphurous hot springs beneath the cave.

Hence, higher life forms can exist even in totally isolated ecosystems like these. The energy sources may vary, but the food chain in these systems ultimately depends on micro-organisms that obtain their energy from grazing on purely inorganic chemicals.

We should now return to the basic question. What are the minimum elementary components that must be present in an environment to make it habitable by living things? From the fact that life forms thrive in extreme environments such as we have described and the fact that intricate ecosystems can exist in pockets of isolation, it is obvious that many of the everyday features of our surroundings that we humans regard as essential for our survival are simply not a necessity for other organisms.

We have seen that oxygen (free oxygen, that is, in the form of an atmospheric gas) is unimportant to certain anaerobes and can even be lethal. We have also seen that a continuing supply of organic food is not essential to the autotrophs, though a source of carbon-based molecules must have been available to allow life to get going in the first place.

Plants need light and most animals need plants because they are the primary source of organic matter on which our familiar world survives. Nevertheless, light is not essential to the chemoautotrophs, nor even to the animals of the Movile Cave.

One of the key common factors in all the environments where life exists is the presence of water. For a start, living cells are composed largely of water. Water is a highly effective solvent and can dissolve a wider variety of molecules than other liquids. It therefore acts as a medium in which the molecules in a cell can move around and interact. This is how the metabolic chemical reactions are able to take place. Water is also the medium by which the waste products can be transported away.

In addition, water is itself an agent in many of these reactions. For example it can easily be broken down into its component elements hydrogen and oxygen, which can be taken up and incorporated into biologically important molecules. It is just about conceivable that some other liquid might be capable of some of these vital functions, perhaps liquid ammonia on a colder planet, but no other known solvent remotely compares to water in its effectiveness.

Let us now identify the essential factors needed in an environment in order to make it habitable. It seems that these can be summarised in the following rather simple list.

A supply of energy. Energy is needed to drive the metabolic processes in any living organism. This need not be solar energy, but could be simple inorganic chemical energy, and this is available from a variety of sources.

A supply of basic chemical elements. These are needed to build molecules like the amino acids and nucleotides as well as the structural components of the cell. Hydrogen, carbon, nitrogen, phosphorous and sulphur are the most important in this respect, though traces of other elements are also needed. These are amongst the most abundant elements in the Universe and all the evidence shows that they are widely distributed throughout it.

A supply of water. This of course means liquid water. Although water is very common in the Universe – it is for example present in vast quantities in the interstellar clouds – liquid water is by no means common. It is normally present either as a gas (as in the interstellar clouds) or as a solid (as on the surfaces of the icy satellites). In fact liquid water can exist only within a rather narrow range of temperatures and pressures, and so this in turn leads to the following requirement.

A temperature and pressure at which water is in its liquid state. This is the most restrictive criterion of all, far more stringent than the requirement of a supply of energy and chemicals.

There is one more environmental requirement that ought to be mentioned, protection from harmful radiation. Various kinds of high-energy radiation can injure living cells by interfering with cell division, by damaging their genetic material or even by killing them outright. In particular, all life forms are vulnerable to a greater or lesser extent to ultraviolet radiation; some terrestrial micro-organisms would be destroyed by exposure to even the small amount of solar ultraviolet light that filters through the Earth's atmosphere. The reason for this vulnerability is that nucleic acids absorb light very effectively at mid-ultraviolet wavelengths, becoming damaged or broken up in the process. Although this radiation is quite easily blocked by water, soil or rock, an unsheltered surface on, for example, a planet with no ozone layer would be a hostile place for life to exist.

Even so, some microbes are known to have an amazing tolerance to radiation. The bacteria *Micrococcus radiophilus* and *Deinococcus radiodurans* have developed very efficient mechanisms for repairing their DNA and can recover from intense X-ray doses that would easily destroy other organisms. And certain Antarctic cyanobacteria exposed to mid-ultraviolet radiation generate their own sunscreen pigments. They produce an outer sheath of mucus containing a substance that filters out the damaging rays. Hence, although protection from radiation is certainly needed, it cannot be considered an absolute requirement of the environment.

Notice that the above list of essentials doesn't include any need for solar (or, in the case of another star, stellar) radiation. On Earth most life depends on solar energy. It powers photosynthesis and, most importantly, maintains a suitable temperature for water to exist as a liquid. However, it seems that, in principle at least, it is not an essential. We could conceive of an ecosystem feeding on inorganic chemicals and warmed only by, say, the geothermal heat of a planet, though such an ecosystem is likely to remain very primitive. All the respiring terrestrial life forms depend on solar energy as well as a supply of free oxygen. These more complex life forms have developed and diversified because they have been able to exploit the greater efficiency of the processes of respiration and photosynthesis.

The importance of water

The need for liquid water is the hardest of all the criteria to satisfy and it looks as if the most productive strategy to adopt in the search for extant life elsewhere in the Universe is likely to be a search for water. The presence of water as a liquid implies a temperature range of about 0 – 100°C, somewhat higher under pressure, together with an atmosphere or other protective layer to prevent it from boiling off into the vacuum of space. This temperature range is a very severe restriction on the range of habitable environments.

5.4 - *An Antarctic dry valley.*

Yet even water may not be needed continuously. One of the driest places on Earth lies deep in the interior of the Antarctic continent. Here, on the high plateau, there are desert valleys that are kept clear of snow and ice by the scouring winds (Fig. 5.4). The environment is exceedingly harsh. Air temperatures are normally well below freezing, rising above zero on only a very few days each year. There is essentially no precipitation and the winds remove all moisture from the air. The valleys are so cold and dry that in these respects the conditions mimic those found on the surface of Mars; in fact this was one of the locations chosen by NASA to test out their Martian rover vehicles.

Despite all this, life is present. Break open one of the sandstone rocks lying around and you will discover bacteria and lichen growing a few millimetres under the surface (Fig. 5.5). The porous rock is able to retain just a trace of water and the rock temperature can rise a little above freezing if warmed by the Sun. Needless to say, the metabolism of these organisms is extremely slow. For the Antarctic bacteria to complete one cycle of absorption, excretion and replacement of carbon takes something like 10,000 years. They are the slowest-growing organisms known.

5.5 - *Micro-organisms living just beneath the surface of porous rocks from the Antarctic dry valleys.*

Many life forms are able to survive entirely without liquid water for very long periods of time by suspending their metabolism altogether and entering a state of dormancy. At the opposite end of the world from Antarctica, at a site near Fairbanks, Alaska, is the Fox Permafrost Research Tunnel. This 110-metre long tunnel was excavated in the permanently frozen silt and gravel on the site of an old goldmine and is used as an underground laboratory for a variety of research programmes. Samples of long-buried mosses have been recovered from the tunnel and thawed out. They started to grow again despite having been completely frozen for a period of time estimated at some 40,000 years.

As in all other respects, the microbes outperform all other living things in their resilience and endurance. It seems that bacteria can remain dormant for indefinite periods and can be revived unharmed as soon as the right conditions are restored. Dormant bacillus spores trapped in Dominican amber have been reanimated after an entombment of at least 25 million years. It seems that bacteria like these can dehydrate themselves and shield their DNA with protective proteins that stop it from deteriorating. When the spores eventually revive, they use specialised enzymes to repair any DNA damage that might have occurred.

Moreover, there are bacteria that may well have survived an incredible 250 million years of dormancy. A bacillus was successfully cultured after being extracted from a crystal that was laid down in a salt bed near Carlsbad in New Mexico during the Permian period. This amazing specimen raises the question as to whether spores like this may, in effect, be immortal.

The habitable zone

These remarkable survival stories do not alter the conclusion that liquid water is essential for the metabolic and developmental processes that drive living organisms. The best strategy for a search for extraterrestrial locations supporting extant life is still the search for liquid water. However, there may also be places that although waterless now, contained significant quantities of water in their earlier histories; a planet, perhaps, with an ocean that has now dried up. These are places where the fossils of extinct life forms might well be preserved. They might even continue to shelter dormant organisms like bacterial spores. Hence, any signs of the former presence of water also merit close attention.

As already stated, for water to be present as a liquid, a temperature range of between 0°C and a little above 100°C is required. The surface of the Earth obviously falls into this range; the average surface temperature is about 15°C. This temperature is maintained almost entirely by energy from the Sun, supplemented to only a very minor extent by internal, geothermal sources. The Earth is also radiating heat energy away into space, and it is the balance of incoming and outgoing radiation that keeps the average temperature steady. The atmosphere plays an important part here, slowing the loss of heat energy and insulating the surface. Without an atmosphere, the surface temperature would be some 33°C cooler, at a distinctly unfavourable −18°C.

Not surprisingly, the closer a planet is to the Sun, the higher its temperature is likely to be. Mercury, the closest, has an average surface temperature of 170°C, whereas the outermost gas giant Neptune is at about −220°C. However, the actual surface temperature depends on other factors too, in particular the nature of the atmospheric cover. In the case of Venus, an extensive insulating atmosphere has pushed it up as high as 460°C, a lot higher than the melting point of lead.

This variation in the surface temperatures of the planets has given rise to the concept of the habitable zone. This is the range of distances from the Sun within which a planet might have a surface temperature that permits water to exist as a liquid. Within the Solar System, only the Earth lies clearly within this zone. Venus lies near the inner (hotter) boundary of this zone and Mars lies near the outer (colder) boundary. The idea of a habitable zone is also applicable to other stars, though the zones will not all be of the same size. Hotter stars will have more extended habitable zones and cooler stars will have narrower ones.

The idea of the habitable zone is useful in as much as it defines one region within which we might hope to find planets with liquid water. However, as we shall see in Chapter 8, even in the Solar System there may well be other places where the temperature is in the right range for water to be present.

In the case of the Earth, the fact that it orbits comfortably within the habitable zone of the Solar System is a key factor in its suitability for life. Its distance from the Sun allows the existence of oceans at its surface. The atmosphere also plays a key role, helping maintain the temperature and preventing the evaporation of the oceans into space. And, in turn, life itself has helped create the oxygen-rich atmosphere that we enjoy today.

This concept of the Earth as an interlocking set of parts, continents, oceans, atmosphere and life, each one constantly changing and at the same time changing the others, is an idea that we shall now examine in more detail.

6 The evolution of Earth and life

Whatever its origin, it is clear that life on Earth has both changed and diversified over the last 3,500 million years. To what extent might we expect extraterrestrial life to have developed in the same way as it has on Earth? Is it likely or even inevitable that wherever life develops, plants and animals should eventually appear, or intelligent beings emerge?

Before we tackle this question we first need to recognise the extent to which life and the Earth have evolved together and how much each has influenced the other's development. The Earth has evolved since its formation very largely because of internal geological forces but also because its own astronomical environment has been in a continual state of flux. Therefore the theme of this chapter is change, evolution and the mutual interaction between life and the Earth.

Biological evolution

Once life of some sort had established itself on the early Earth, it developed according to the principles of biological evolution. Charles Darwin was the first to provide a scientific explanation of how evolution takes place. In 1859 he published *On the Origin of Species by Means of Natural Selection*, in which he established the theory and explained the fundamental concept of natural selection that underpins it.

Although this theory accounts for the diversity of life that we observe around us and its development from a common ancestor, it is not very helpful in resolving the problem of the origins of life itself. It is far easier to comprehend the course of evolution since the time of the first appearance of bacteria in the geological record than it is to understand the events that led up to their emergence.

Biological evolution is a process of descent in which there is gradual change between successive generations. One kind of change comes from sexual reproduction, children are different from their parents. Another source is mutations, the minute alterations that take place in an organism's genetic material because of errors in replication or exposure to high-energy radiation or toxic chemicals.

A mutation, for example, could cause malfunction of the organism or even death. Nevertheless, sometimes the organism continues to work without ill effects. The mutation could even confer an advantage under certain circumstances; that is, in the appropriate environment. Neutral or favourable mutations will be carried to the next generation.

These changes become the cause of random diversity within a population and natural selection acts on diversity. In a competition for resources, the survival of an individual depends in a broad sense on its suitability for its environment, including its ability to compete for the protection it affords and the sources of food available. If the environment changes or the population migrates to a new environment, natural selection tends to operate so that the best adapted organisms have a better chance of survival and reproduction and the least adapted are more likely to be eliminated. The survivors are not (necessarily) the biggest or the strongest, simply the most suitable for the new environment. In this way, change gradually accumulates step by step from generation to generation. Although the diversification is random, the selection process is most definitely not.

Darwin identified the three key stages of the evolutionary process, namely variation, replication and elimination by competition but it was not until the twentieth century that the science of genetics revealed in detail how these processes work, leading to the modern understanding of the theory.

Nothing in this theory predicts the progression of the species that emerge. The evolutionary rules do not inevitably dictate, to take an example, the eventual emergence of animals, mammals or apes. In no sense is there a pre-determined evolutionary ladder to climb. However, there is a tendency for life to become more complex and more diverse as time goes on. Multi-celled animals with specialised organs are certainly more complex than single-celled prokaryotes, both in structure and operation.

Hence there is nothing inevitable about the emergence of a particular species or even a particular kingdom, domain or attribute. We human beings pride ourselves on being something special in the animal world, and credit ourselves with a quality that we have called intelligence. We recognise this property in other animals, though we have tended to believe it is rather less developed in them. This raises the question: is intelligence inevitable? Is it an attribute that, given time, will always arise out of this tendency towards increasing complexity? Is it, in fact, even useful in evolutionary terms? We will return to these issues in Chapter 10.

Of course, simpler organisms continue to coexist alongside more complex ones. Humans did not evolve from modern apes, but humans and modern apes share a relatively recent common ancestor, a species that is now extinct, and in purely biological terms it would be impossible to say whether or not a human was more complex than an ape. However, humans, amoebae and indeed all the living things on Earth also share a common ancestor. Yet it is not difficult to tell that humans are more complex than amoebae.

Environmental change

If genetics is the basis of variation and reproduction, then much of the impetus behind natural selection comes from environmental change. Clearly, environmental conditions at the Earth's surface have altered in many respects from, say,

6.1 - The pattern of the continents as they were 160 million (upper) and 100 million (lower) years ago.

the Hadean era. Continents now cover a third of the Earth's surface, great mountain ranges have been thrown up and the atmosphere has become rich in oxygen. Plants and animals have colonised both continents and oceans.

6.1 - The pattern of the continents as they were 160 million (upper) and 100 million (lower) years ago.

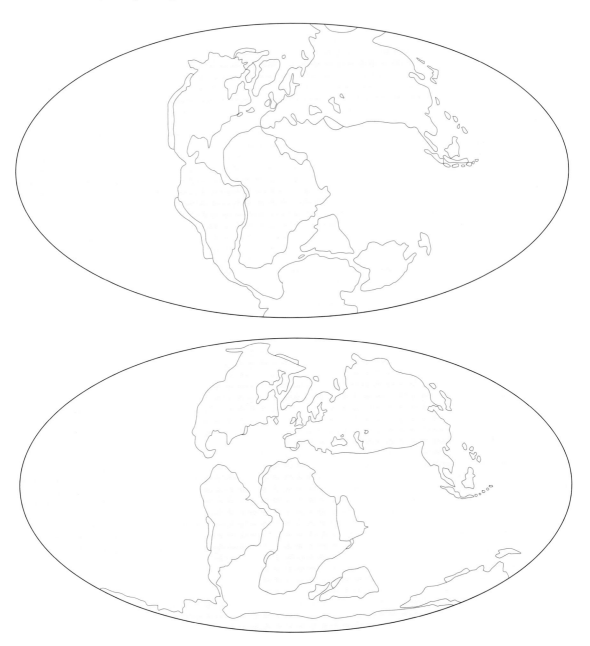

These changes did not take place in a smooth and continuous way. The intervening time has been one of continuous fluctuation. The geological forces that raised up

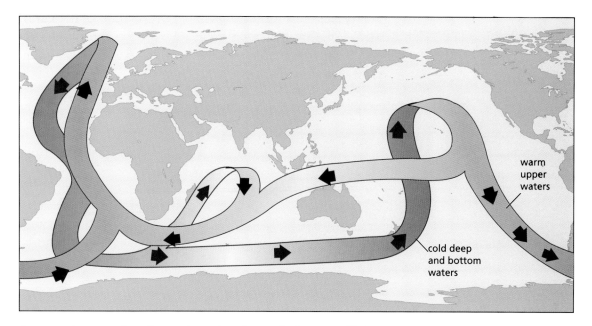

warm
upper
waters

cold deep
and bottom
waters

6.2 - *The deep oceanic circulation system that links the Atlantic, Indian and Pacific Oceans.*

the continents also formed and reformed them, splitting them apart, pushing them together again and constantly moving them around the globe (Fig. 6.1). Oceanic circulation patterns were established and then interrupted again by these movements. Regions of snow and ice formed on polar continents and on high mountain ranges but then disappeared again as the continents moved back to warmer regions or the mountains weathered away. All these physical upheavals were in turn the causes of great climatic changes.

The Earth's equatorial regions receive more heat from the Sun than the poles. This heat is distributed around the planet by great systems of winds and currents; if this did not happen, the differences between tropical and polar temperatures would be far more extreme than they actually are. For example, the Gulf Stream is a fast-flowing oceanic current that carries warm water from the east coast of North America into the North Atlantic west of the British Isles. Here much of its heat is given up to the air and carried by the prevailing winds across Western Europe, maintaining a warmer climate than its latitude would otherwise allow. The Gulf Stream is part of a surface current system that circulates round the entire North Atlantic but it also connects with a deep circulation system that extends across the whole globe, linking the Atlantic, Indian and Pacific Oceans (Fig. 6.2).

These great circulatory systems are long-established, though not permanent; there is evidence, for example, that the North Atlantic deep water formation may even have shut down temporarily about 11,000 years ago, possibly causing an increase in glacial activity and colder climatic conditions in northwestern Europe. These systems are driven by solar heating, but the actual flow pattern they follow is determined largely by the distribution of the oceans and continents.

Over geological timescales of say, a few million years, oceanic circulatory systems like these become disrupted by the movements of the continents. Patterns establish themselves, then break up again only for new ones to emerge. Significant changes to the climate can result, altering the global temperature and weather patterns.

Other long-term geological forces are also at work. To take one example, there may have been sustained periods of volcanic activity, perhaps lasting millions of years, in which vast quantities of greenhouse gases were released into the atmosphere, resulting in natural global warming and extended periods of increasing temperatures. In fact the Earth's climate at any one time in history is the result of the highly complex interaction of these and many other dynamic components, geological, oceanic, atmospheric and biological. Moreover, these factors themselves change, and continue to change, on a number of different timescales.

As well as the climatic influences driven by geological activity, there are also external factors, arising outside the Earth. The most important of these are associated with changes in the amount of energy arriving from the Sun.

Solar energy is the force that drives the wind systems and the oceanic currents. Yet the energy output of the Sun is not entirely constant. It has been long known that the Sun undergoes regular periodic increases of activity that are related to the cyclical growth and decline of its magnetic field. At periods of high activity, which occur on an eleven-year cycle, it is prone to throwing off material from its surface in outbursts called coronal mass ejections. These can create spectacular displays of aurorae in the polar regions of the Earth – the so-called northern or southern lights – and cause considerable disruption to terrestrial and satellite radio communications.

The brightness of the Sun varies, very slightly, in phase with this eleven-year cycle. The fluctuations are small (at the peak of its cycle the Sun is only a fraction of a percent brighter) and their eleven-year timescale is so short that it was at first thought unlikely that they could have any detectable effects on the climate. However, the intensity of this activity also seems to fluctuate over periods of about 80 to 90 years (and there may be other, longer timescales too) and there is now good evidence that periods of increased solar activity correspond to periods of slightly higher global temperatures by perhaps a few tenths of a degree. Although the evidence is convincing, the exact way in which this comes about is poorly understood.

This is a good illustration of the fact that when we are considering the effects of a change in one of the factors that affect the climate, not only the size of the change but also the timescale on which it occurs is important. Some components of the climate, for example the volume of ice in polar icecaps and glaciers, may take thousands of years to respond.

Astronomers have made detailed calculations of the changes in the rate at which the Sun consumes its hydrogen fuel over the very long term, and these show that it has brightened steadily by as much as 30% since the Earth was formed. Of course, from year to year, or even from century to century, this change is undetectably small, but the long-term effects of a difference of this amount could be expected to be very significant.

The Milankovich cycles

Some of the most important long-term external influences on the Earth's climate result from cyclical changes related to the Earth's orbit round the Sun. These astronomical cycles

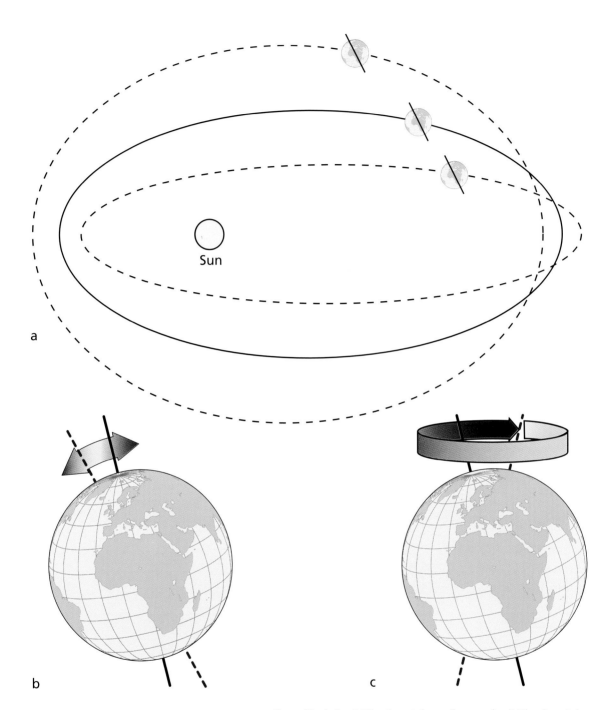

6.3 - *The three Milankovich cycles (a) an 110,000-year cycle in the eccentricity of the Earth's orbit, (b) a 40,000-year cycle in the tilt angle of the Earth's axis and (c) a 22,000-year cycle in the direction of the Earth's axis.*

are usually called the Milankovich cycles or the Milankovich-Croll cycles. James Croll, a Scotsman working in the 1860s, suggested that these orbital rhythms might affect the Earth's climate and the Serbian astronomer Milutin Milankovich went on to develop the theory in some detail in the 1920s. There are three of these cycles (Fig. 6.3).

EARTH, LIFE AND THE UNIVERSE

The first occurs because of changes in the elongation of the Earth's orbit. The orbit is not completely circular, but has the form of an ellipse in which the Earth is slightly nearer to the Sun during one part of the orbit (perihelion) than at the other (aphelion). It is about 147 million kilometres away at perihelion and about 153 million kilometres at aphelion. The degree of elongation of the orbit is called its eccentricity, and because of the gravitational effects of the Sun and planets, this eccentricity varies with time, taking about 110,000 years to change from a nearly circular shape to a more elliptical one and back again. When the orbit is more elongated, the seasons in one hemisphere (the hemisphere in which summer occurs when the Earth is closest to the Sun) are very exaggerated whereas in the other hemisphere they are quite moderate.

The other two Milankovich cycles are caused by changes in the tilt of the Earth's axis. This angle is currently about 23.4°, but it oscillates between about 21.8° and 24.4° on a cycle that lasts about 40,000 years. When the tilt is greatest, the contrast between winter and summer is increased – winters are colder and summers are hotter.

The direction in which the Earth's axis is tilted is also changing. The Earth's axis slowly wobbles like a spinning top in a cycle called precession that takes about 22,000 years to complete. This is the shortest of the cycles and it determines which hemisphere has its summer when the Earth is at perihelion. At the moment this is the southern hemisphere, but in 11,000 years time it will be the northern hemisphere.

change in	period (years)	effect
the eccentricity of the Earth's orbit	110,000	determines which hemisphere has the more extreme seasonal effects
the tilt angle of the Earth's axis	40,000	affects the contrast between winter and summer
the direction of the Earth's axis (precession)	22,000	determines which hemisphere has its summer when the Earth is at perihelion

Each of the effects is small, but, particularly when they reinforce one another, they can be enough to make a significant difference to both the annual amount of solar radiation reaching the Earth and the distribution of solar heating between latitudes or hemispheres. These effects can become further amplified if, as they are at present, the continental masses are concentrated at particular latitudes, since continents and oceans respond differently to solar heating. Land masses heat up and cool down relatively rapidly, but oceans retain heat better and react more slowly to changes in incoming solar radiation.

As an illustration of the way in which these sorts of changes can affect the Earth's climate, consider the effect of the contrast between the seasons on the size of the polar ice caps. In a period when the seasonal variation is reduced, so that the summers are cooler, there will be less summer thawing and the ice sheets will tend to grow. Greater seasonal variation with hotter summers causes more summer melting and the ice sheets will shrink.

The ice caps originate from water evaporated from the oceans, which falls as snow at high latitudes. In a phase in which the ice caps grow, more and more water is removed from the oceans and the sea level falls as the ice volume increases.

The oxygen component of the water that evaporates from the oceans is preferentially taken up in the form of the isotope oxygen-16 rather than oxygen-18. Hence, in any period in which the ice caps are growing, the polar ice becomes enriched in oxygen-16 but the oceans become richer in oxygen-18. Therefore, by measuring the oxygen isotope ratios in ice cores drilled from the polar ice sheets (Fig. 5.1) it is possible to monitor the course of this process over hundreds of thousands of years, recording changes in temperatures, ice volumes and sea levels. It is possible to push the record back even further by looking at oxygen isotopes in sea-floor sediments.

The results show a good correlation with the variations in the amount of solar radiation that were predicted by Milankovich. In fact it was found that the Milankovich cycles could explain most of the observed temperature fluctuation, and there is evidence that they may have been

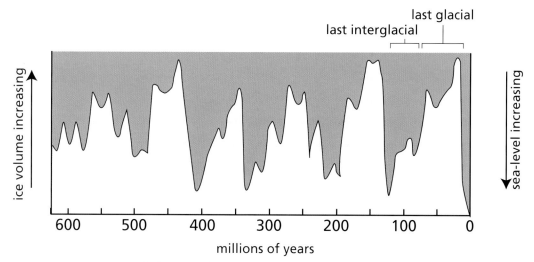

6.4 - *The cycle of glacial and interglacial periods within the present Ice Age.*

important instruments of climatic change for a period of at least two million years. During this time, global mean temperatures have probably fluctuated by 5°C or more and sea levels by several tens of metres.

These periods of advance and retreat of the ice sheets covering northern Europe and North America are called glacials and interglacials. They are presently occurring on a timescale of about 100,000 years, corresponding to the longest of the Milankovich cycles (Fig. 6.4). These large ice sheets first appeared about two million years ago, marking the start of the present Ice Age in which the Earth still finds itself today.

Ice Ages

Ice Ages are periods lasting several million years when extensive areas of ice cover the Earth (but note that the term Ice Age is sometimes also rather confusingly used to refer to the much shorter glacial periods – see the glossary for more about this). Major Ice Ages have occurred a number of times in the past; for example towards the end of the Precambrian, between 700 and 800 million years ago, and again about 280 million years ago at the boundary of the Carboniferous and Permian periods. However, the Ice Ages are relatively rare periods of geological history. During the 90% of the time that the Earth has not been in an Ice Age, mean global temperatures were several degrees higher than at present.

Although Milankovich cycles can explain climate change within an Ice Age, they cannot explain what triggers the occurrence of an Ice Age in the first place. Some important clues to the causes of Ice Ages were uncovered in the 1990s by the climatologist Robert Berner of Yale University. His calculation took into account the rate at which carbon dioxide (CO_2) enters and leaves the atmosphere as a result of processes like the weathering of the continents, volcanism and biological activity. He was then able to track the changes in the atmospheric concentration of CO_2 over hundreds of millions of years. His work strongly suggested that the level of carbon dioxide is higher during warm periods in the Earth's history and lower during the Ice Ages. Carbon dioxide is the most important of the so-called greenhouse gases, and so it seems very likely that these gases play a significant role in regulating the Earth's long-term climate.

As well as carbon dioxide, the greenhouse gases also include water vapour, ozone and methane. Their presence in the atmosphere is in fact essential for life today, since without them the temperature at the Earth's surface would be about 30°C colder than it actually is. These gases are transparent to visible radiation but not to infrared radiation. Although they admit most of the incoming visible radiation of the Sun, they block and absorb much of the outgoing infrared radiation from the surface of the Earth, with the result that the atmosphere warms up. This is called the greenhouse effect because it works rather like a greenhouse, keeping the temperature up by creating an insulating blanket of warm air.

The amount of greenhouse heating depends on the concentration of these gases in the atmosphere. If the concentration rises or falls, so does the temperature at the surface of the Earth. The Swedish chemist and Nobel laureate Svante Arrhenius in 1896 was the first to suggest that carbon dioxide could affect the planetary temperature in this way. He also predicted that the burning of fossil fuels would cause carbon dioxide to accumulate in the atmosphere and induce a global temperature rise.

The massive increase of CO_2 and other emissions caused by human activity in the last 150 years, particularly through fossil fuel burning, is now widely believed to be the cause of present-day global warming. Today's CO_2 levels, 360 parts per million, are higher than any recorded in the Vostok ice cores over the past 400,000 years. The increase (up from 280 parts per million in the middle of the nineteenth century) is taking place some 200 times faster than it did during the most recent interglacial warming. In 2001, the Intergovernmental Panel on Climate Change predicted that by the year 2100 the mean global surface temperature would rise by a further 2.0 – 4.5°C and by up to 6°C over some continental land masses.

Given the complexity of the interactions between the different components of the Earth's climate, it is unlikely that any single factor would have been responsible for initiating an Ice Age. However, it is interesting that biological triggers have been suggested in at least two instances. Living organisms, especially plants, extract carbon dioxide from the air, and the Ice Age near the end of the Precambrian could perhaps have resulted from global cooling caused by the explosive increase of single-celled organisms and the relatively sudden removal of large amounts of CO_2 from the atmosphere.

Similarly, the late Carboniferous and early Permian Ice Age coincided with the spread of vast regions of swamp. These became buried by sediment (eventually forming coal seams), removing huge amounts of carbon from the surface. This carbon, in the form of carbon dioxide, had originally been extracted from the air by plants as they grew and formed the swamps.

Nobody can be sure whether this kind of fall in the atmospheric carbon dioxide level would by itself have reduced global temperatures far enough to provoke an Ice Age, since the impact of living things on the climate is inextricably bound up with geological and other factors in a complex web of events. Nevertheless, whether or not life was a critical part of the process, there is no doubt that it played a role in the development of the climate.

Evolution of Earth and life

It is not hard to see how these climatic and environmental changes would have helped drive the course of evolution of life on Earth. Changes in air or sea temperature or the composition of the atmosphere would have favoured one evolutionary path over another. Extensive ice coverage would have caused extinction of some species or forced the migration of others to new environments. Sea level changes would have exposed continental shelves or covered coastal plains, providing new habitats or annihilating older ones.

The separation of continents or the appearance of mountain ranges might physically divide members of a species so that afterwards they would diverge in their evolution. For example, the fossil record tells us that the marsupials such as the kangaroos, opossums and koalas that now inhabit Australasia and South and Central America actually originated in the New World. It is believed that they crossed to Australia by land some time before it became isolated from the other continents, an event that took place perhaps about 40 million years ago. Here, in the absence of competition from the rival placental mammals, the marsupials were able to diversify widely and successfully.

In ways like this, the evolution of life was driven by the evolution of the Earth. And, as we have seen, the reverse is also true. There would for example be essentially no free oxygen, nor ozone, in the atmosphere without life. Therefore, because of its effect on the levels of carbon dioxide in the atmosphere and hence on the greenhouse effect, life plays a part in regulating the Earth's surface temperature. In summary, the Earth, in which we must now include the geosphere, atmosphere, hydrosphere and biosphere, is really a single system of interconnected forces and effects acting over a wide range of timescales.

The links between the component parts of the system are not always easy to understand, but it is helpful to see the way in which certain important constituents are exchanged and cycled between them. Water, for example, moves between the oceans, the atmosphere and the land in a cycle of evaporation, precipitation and run-off and is linked to other surface cycles of weathering, erosion and sediment deposition. In fact there is a whole range of similar phenomena that constantly redistribute chemical elements between different parts of the Earth system. The most important of these is the carbon cycle.

The carbon cycle

The exchange of carbon between plants and the atmosphere is an example of the way in which a chemical element can move around in the Earth system. Plants continually absorb carbon from the air in the form of carbon dioxide (CO_2) during photosynthesis, and store it away in the form of carbohydrates. What happens to this carbon? Some of it is eaten by animals, though this is actually an insignificant amount of the total. In fact about half of it is returned to the air (again as CO_2) through the process of respiration and about half ends up in the soil as organic matter when the plant dies.

We can picture carbon as being temporarily stored in a number of locations or reservoirs like the atmosphere, the plants and the soil, with the carbon being constantly moved between them by various means. The process of photosynthesis and respiration described above is not a closed cycle, because the net effect is to extract carbon from the air and deposit it in the soil. In fact much of the carbon in the soil returns directly to the air when the plant debris decomposes, completing the cycle, though a certain amount is washed out into streams and rivers, eventually ending up in the oceans, which are another reservoir.

There are actually five main surface carbon reservoirs. In order of capacity, they are the oceans, the deep-sea sediments, the soil, the atmosphere and the land plants, which in effect means the forests. The oceanic reservoir, which contains about 90% of the surface carbon, is really made up of a number of subsystems, with carbon being exchanged between deep and surface layers and between the water and the marine organisms that live in it.

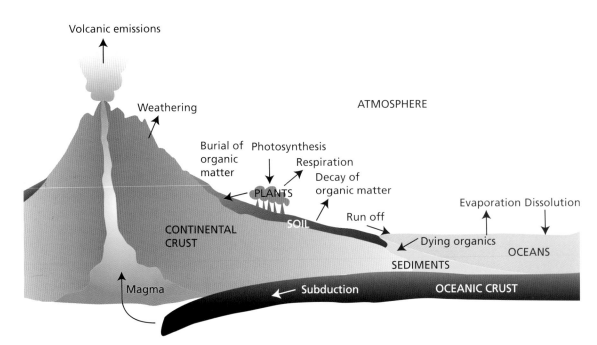

6.5 - *The carbon cycle. Carbon is transferred from one reservoir to another by a variety of mechanisms all proceeding at different rates. The carbon occurs in various forms, for example in the atmosphere mostly as carbon dioxide and in the oceans as carbonates.*

Carbon is circulated between all these reservoirs. For example, carbon dioxide from the atmosphere becomes dissolved in seawater, forming carbonates, and at the same time carbon dioxide in the water evaporates back into the air. Marine organisms take up dissolved carbon in the form of carbonates or carbon dioxide, and this becomes deposited in the seabed sediment when they die.

Overall, these various pathways and cycles maintain a broad balance between the different parts of the Earth system. However, climatic and other conditions can affect the rates at which carbon moves between them, and the carbon content of some reservoirs can grow or decline over time, the increase of carbon dioxide in the atmosphere today being an example.

The entire cycle is really made up of a number of interlocking sub-cycles, each of which proceeds at a different rate. The shortest of these is the terrestrial cycle that operates through the plants and the soil – the average residence time of carbon in plants is just a few months or years. By contrast the marine cycle, measured by the average time it takes for carbon to circulate in and out of the oceans, takes perhaps hundreds or thousands of years. There is however a further important, very long-term component of the carbon cycle.

At first sight, it seems that there is no exit from the carbon reservoir formed by the deep-sea sediments. The shells, skeletons and organic remains of dead marine creatures are constantly adding to this reservoir as they rain down onto the ocean floor. In time, because of the continually increasing pressure of the overlying sediment, chemical changes occur in the deeper layers, and they become lithified, that is, they are converted into rock.

6.6 - Chalk cliffs, lithified carbon deposits that have been brought to the surface and exposed by erosion.

Deposits that have formed mainly from shells or skeletons may become limestone or chalk, and those that come from organic debris may turn into coal seams or oil-rich shales.

It seems as if lithification is a way of removing carbon from the cycle altogether, but this is true only in the short term. The rocks, along with the rest of the outer layers of the Earth's surface, are in constant movement, subject to slowly-acting but inevitable geological forces. Buried rocks can be brought to the surface and exposed (Fig. 6.6). The uplifted rocks are weathered by wind and water, releasing carbon dioxide back to the atmosphere or washing organic carbon into the rivers and seas.

At the same time, in a different process, the ocean floor sediments can be pulled down into the mantle beneath the crust at continental boundaries in the process called subduction. In the higher temperatures of the Earth's interior, the lithified rocks release their carbon dioxide together with water and other volatiles in a mixture that causes the mantle rocks to melt into magma. When the magma reaches the surface, volcanoes form and vent carbon dioxide and other gases back into the atmosphere, finally closing the loop.

The geological cycle is of course by far the slowest of all those that together make up the carbon cycle. For a carbon atom to pass through all the stages of sedimentary deposit, subduction, weathering and volcanism probably takes about 100 million years.

The Earth system and the Gaia hypothesis

The operation of the carbon and other cycles shows how a change in any one part of the Earth system might affect any other part, even though the consequences may take a very long time to work through a prolonged chain of events. In principle, an extended volcanic outburst on one side of the globe could result, through more carbon dioxide being added to the atmosphere, increased global temperatures, a melting of the ice caps and a rise in sea level, in the extinction of a marine species living on a continental shelf on the other side of the world. In practice, because of the multiplicity of the interactions between the component parts, no single event is likely to set off a chain of events quite as directly as this unless it is on a truly massive scale.

It can sometimes happen that a shift in the climate produces effects that are self-reinforcing. For example, a fall in global temperature might result in an expansion in snow and ice coverage, both at the poles and on mountain ranges. This would increase the reflectivity of the Earth's surface so that it would absorb less of the incoming solar radiation and lead to an even further drop in temperature. This is an example of positive feedback. Positive feedbacks in the Earth system are destabilising influences on the climate.

There have of course been global temperature swings in the Earth's history. In the depths of the Ice Age glacials, overall temperatures were several degrees lower than today, whereas in the Cretaceous period, they could have been as much as 15 degrees higher, with forests growing at the poles. Yet these are relatively modest variations. Most aspects of the climate, including the temperature, have fluctuated only within quite moderate limits and in this sense they have actually remained comparatively steady for extremely long periods of time.

The action of the carbon cycle is an example of one of a number of mechanisms that contributes to this long-term stability. Imagine that the level of CO_2 in the atmosphere were suddenly enhanced, perhaps by a burst of volcanic activity. As we have seen, this would be expected to result in an increased greenhouse effect and a consequent rise in global temperatures. One knock-on effect might be an increased evaporation rate of water from the seas and soil. This would boost the hydrological cycle and so rainfall would increase.

The resulting warmer and wetter conditions would be expected to stimulate plant growth, which would extract CO_2 from the atmosphere. The pace of weathering and run-off would also increase, resulting in an increased rate of sedimentation and carbon burial in the oceans, in other words a long-term removal of carbon from the surface to compensate for the original emissions. In this case the feedback is negative; the original effect is counteracted, given enough time. Negative feedback loops like these are forces for stability in the Earth's climatic system.

Considering the variety of the forces at work and the complexity of the interactions between them, it is perhaps surprising that the surface temperature of the Earth has fluctuated so little since the Hadean. The oceans have been in continuous existence since then, with the implication that the global mean temperature has never deviated much outside the range that keeps them liquid. This is even more remarkable considering that the energy input from the Sun was much less in the distant past, and has increased by as much as 30% since the Earth was formed.

Other things being equal, this would have meant that Hadean surface temperatures would have been some 20°C lower than today's, enough to freeze the oceans completely and make the emergence of life much less likely. Yet the geological evidence shows that the oceans were liquid. This is known as the 'weak Sun paradox'. The explanation for this is not entirely clear, but it is possible that the greenhouse effect may have been much greater at that time. Analysis of ancient soils and rocks shows that the air contained far too little carbon dioxide to have kept the planet from freezing, but another greenhouse gas, methane, was probably more abundant and may have had the necessary effect. Methane is a product of many simple prokaryotic life forms and methane-generating organisms would have been common at that time.

In complicated systems like the Earth system, with many components and a multitude of forces linked by both positive and negative feedback loops, there is a certain built-in robustness and stability, with a tendency to settle into one of a number of steady states. If one part of the system (say the CO_2 content of the atmosphere) becomes unbalanced for some reason, it will have a knock-on effect on other components, launching a chain of disturbances that propagate their way through the rest of the system. The net result, after a certain response-time delay, is usually to settle back close to its original state. The Earth system seems to be in a stable state like this, where the mean global temperature has not strayed outside the 0 – 30°C range.

Living things, particularly plants, play an indispensable part in this system. They extract carbon dioxide from the atmosphere, acting as a thermostat that keeps the Earth temperature steady by moderating the greenhouse effect. In fact, life plays a key role in the continued existence of water on the surface of the Earth, neatly closing the circle, since water in turn is essential to life.

In 1972, observing the fact that the Earth and the life on it have evolved together in a closely coupled system in which certain key features such as the long-term maintenance of surface water appear to be self-regulated, the British scientist James Lovelock framed the Gaia hypothesis. In brief, this asserts that the Earth's surface environment, its climate, temperature and chemical composition is and has been maintained in a state that is favourable for the support of living things. This self-regulation is not purposeful, but automatic. In this conception, the Earth is very like a kind of super-organism.

At first sight, there seems to be no reason to expect that the complex web of feedbacks within the Earth system should work in such a way as to regulate it to be conducive to life. Nor is there any obvious support from the theory of natural selection, which would simply suggest that if conditions changed, life would either adapt to the new environment or die out. However, Lovelock was able to construct a series of simple models that showed how just such a self-regulating system can emerge from a world that includes biological entities that all obey the normal rules of Darwinian natural selection.

Indeed, living things play an essential part in the preservation of the Earth system. Having emerged on the Earth at a time when the environment was favourable for the development of life, life has helped to maintain it in that state ever since. No internal or external disruption has so far been large enough to upset it. Yet there is no guarantee that this stability is permanent. If a disruption in one of its component parts became too large, the system could be pushed beyond what it is capable of adjusting to. It could become unstable, and might settle down again only when it had reached a new, far harsher state.

This potentially catastrophic response of the Earth system is one of the underlying worries about the present-day massive increases in CO_2 emissions. Even in the short term, that is to say on the timescale of decades, ever-increasing levels of carbon dioxide in the atmosphere will result in higher global temperatures that are likely to have damaging effects on the weather, farming patterns and sea levels. Moreover, in the longer term there is also the danger that the Earth might enter a phase of disastrous instability. As the Earth heats up, its biosphere may start to hasten rather than slow global warming. As the temperature rises, the amount of CO_2 that can be absorbed by plants reaches saturation point whereas the amount generated by micro-organisms in the soil continues to increase. What was previously a negative feedback now becomes a positive one and the temperature rise accelerates. This is the runaway greenhouse effect. Eventually a new, hotter equilibrium might be established. But, who knows whether this will be at 50°C, 100°C, or even higher?

7 The mass extinctions

In 1783 an event took place in Iceland that was eventually to lead on to one of the most widespread environmental disasters of recent times. On June 8, lava started to flow from a 30 kilometre-long fissure on Mount Laki, just south of the Vatna Glacier. The progress of the eruption was well documented: lava continued to pour out until the following February, by which time it had covered an area of hundreds of square kilometres and filled two river valleys. About 14 cubic kilometres of lava were emitted, the greatest volume measured in modern history. Events like these are called basalt lava eruptions (Fig. 7.1).

Huge quantities of volcanic gases including sulphur dioxide and hydrogen fluoride were discharged into the atmosphere, forming a low-lying haze of acid aerosols. These gases poisoned and destroyed most of the summer crops and the 'haze famine' that followed led to the deaths of three-quarters of Iceland's livestock and a quarter of its population.

The haze spread widely, dimming the Sun over most of continental Europe, extending as far as northern Africa and western Siberia and causing disastrous harvest failures in Scandinavia. The winter of 1783-84 was abnormally cold over the whole of the northern hemisphere and the lowest ever winter temperature was recorded in the eastern United States, some 5°C below normal. The overall cooling of the northern hemisphere was about 1°C, but the effect on regional weather was very mixed, even causing exceptionally high rainfall in Japan where there were widespread crop losses and famine.

Basalt lava eruptions like this are rare, but their consequences are widespread. Yet the Laki fissure pales into insignificance when considered alongside other such eruptions in geological history.

About 250 million years ago, in their perpetual wanderings across the globe, the Earth's landmasses had by chance formed themselves into a single supercontinent called Pangea that in total length extended from the north to the south pole. The rest of the planet's surface at that time was one huge ocean. This was a very hot period of Earth's history. Yet there was already an abundance of life both on the land and in the sea. Early reptiles and amphibians had long since colonised the land, and there were well-established forests in cooler, high-latitude regions.

7.1 - *A modern basalt lava eruption from the Hawaiian volcano Mauna Loa.*

EARTH, LIFE AND THE UNIVERSE

On this supercontinent, in a region that today forms part of northern Siberia, a truly massive basalt lava eruption took place. It lasted a million years and buried an area of two and a half million square kilometres, an area comparable to the whole of western Europe including France, Spain, Germany, Italy and the British Isles, to a depth of up to three kilometres.

The time of the eruption coincided with the disappearance of enormous numbers of species from the face of the Earth. Over a period of some three to eight million years, about 90% of the large animal species became extinct, vanishing from the later geological record. Marine organisms suffered the most. Perhaps 95% of the oceanic species disappeared, including the trilobites, multi-segmented animals that had been prevalent and abundant in the seas for nearly 300 million years. It was the largest extinction event ever.

The coincidence of the eruption and the extinction cannot be ignored. Huge amounts of volcanic gases and dust must have been vented into the atmosphere. The sulphurous gases would have reacted with the water vapour and precipitated out, deluging the surface with acid rain. The increase of carbon dioxide in the atmosphere would pose the threat of long term global warming. On the other hand, vast amounts of volcanic dust would have had the opposite effect, blanketing the Earth and attenuating the radiation from the Sun. It is hard to be sure exactly what the repercussions were, though it is certain that the environmental consequences would have been dramatic.

Even so, it is doubtful whether we can attribute the extinctions solely to the Siberian eruption. For one thing, they did not happen in a single catastrophic event but were spread out over a period that lasted much longer than the eruption. Longer-term climatic changes were also under way at the time. For reasons that are unclear, sea levels had fallen to nearly their lowest levels ever, and this in itself would have devastated marine life because of the loss of the continental shelf habitat. And the existence of a single north-south supercontinent would almost certainly have set up unusual atmospheric and oceanic circulation patterns. All these factors, and others, are likely to have played a part in the extinction, with the eruption perhaps being the final straw.

Whatever its cause, this event led to a major change of direction in the course of evolution. As well as representing the extinction of the great majority of existing species on the planet, it also marked the start of the age of the reptiles, including the dinosaurs. In geological terms it signalled the end of the Permian period and the start of the Triassic, separating two great eras, the Paleozoic (ancient life) and the Mesozoic (middle life).

Evolution, extinction and diversification

On the timescale of our own lives, Darwinian evolution appears a very slow process. The evolutionary changes from generation to generation are quite imperceptible to us under normal circumstances. Indeed, until the start of the nineteenth century it would never have occurred to most people that the species were anything other than fixed and unchanging. At about this time, the French naturalist Jean-Baptiste Lamarck proposed the idea that living organisms had evolved over the course of time from lower to higher forms. He believed that human beings were the highest form of life and were the end product of this progression. His was, in fact, the first clearly elucidated theory of evolution, though his ideas were eventually shown to be flawed and they became replaced by Darwin's theory.

We know today that all present-day life forms are cousins, the product of evolution from a common ancestor, and that life has not only evolved but also that it has diversified. It is not simply that there is a large number of different species but ordinarily, there is a continually increasing number of species. This diversity arises naturally as a result of benign or advantageous genetic mutation being passed from one generation to another, and this is a slow process. The process of natural selection, the survival of the best-adapted species, which is inevitably complemented by the process of extinction, the weeding out of the least well adapted is also ordinarily slow.

These observations are supported by the fossil record, represented by the tree of life. As I said before, 99.9% of the species that have existed on Earth have become extinct; the typical mammal species lasts a mere million years. Despite this, the total numbers of different species in existence at any one time has greatly increased and life on Earth is probably more diverse now than it has been at any earlier time.

7.2 - Graph showing how the number of families of marine animals has varied with time, illustrating periods of mass extinction.

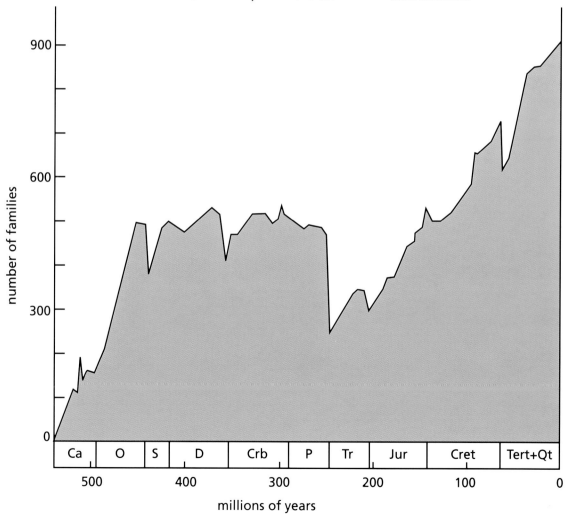

The expansion in the numbers of species is not always steady. Sometimes there are sudden spurts in the numbers and at other times there are declines (Fig. 7.2). Occasionally, as at the end of the Permian period, we observe the steady growth abruptly interrupted by a wholesale loss of species; events like this are known as mass extinctions. These often seem to be followed by a burst of replacement in which an even greater variety of new creatures appears.

The mass extinctions

The five large-scale 'recent' (since the Cambrian explosion) mass extinction events. For a chart showing the geological periods, see page 203.

Five large-scale mass extinction events are recognised in the last 500 million years, that is, since Cambrian times, and these are listed in the table. The dates given are subject to uncertainties of about a million years, a little more in the case of the earlier events.

	millions of years before present	
late Ordovician	443	about 85% of all species lost
late Devonian	354	many marine species lost
late Permian	250	about 95% of all species lost including trilobites
late Triassic	204	about 75% of all species lost, mostly marine
late Cretaceous	204	at least 75% of all species lost including ammonites, dinosaurs

The first of these mass extinctions took place at the end of the Ordovician period, approximately 443 million years ago. At this time, the supercontinent of Gondwanaland, made up from a number of southern continents, lay across the south pole. Extensive ice sheets formed and expanded towards the equator, and an Ice Age began. Sea levels fell and coastal communities were destroyed. As with all mass extinctions, it is impossible to pin down the exact causes, but the devastation to life was probably second only to the end-Permian extinction in its effects, with vast numbers of oceanic species wiped out in about a million years. This catastrophe brought to an end the geological period known as the Ordovician and ushered in the beginning of the Silurian. At the same time, it also marked the start of a new phase of life, its expansion on to the land. Before, only simple plants had made the transition from the oceans, but thereafter, insects, animals and other flora and fauna appeared on the land and colonised its surface.

7.3 - An ammonite fossil. Ammonites had a hollow coiled chambered shell. They first appeared in the Devonian period and became extinct at the end of the Cretaceous.

However, the most recent and probably the most notorious mass extinction took place about 65 million years ago. This is most famously known as the event that resulted in the extinction of the dinosaurs, though its effects were far more comprehensive than this. In a period lasting less than a million years, at least 75% of all land and sea species became extinct. All large land animals including the dinosaurs were lost, as were the flying-reptile pterosaurs and the sea-reptile plesiosaurs. Many marine organisms became extinct, including the ammonites (Fig. 7.3), spiral-shelled animals related to the modern nautilus that had been in existence in one form or another for 340 million years.

The effects on amphibians and the smaller land animals were not quite so drastic. For some reason, turtles, crocodiles, lizards and snakes were affected only slightly, and the mammals, too, mostly survived the devastation. The mammals had appeared on the Earth at roughly the same time as the dinosaurs but had evolved in quite a different way, remaining a subordinate population of small burrowing animals. As had happened in the earlier mass extinctions, the population recovered, taking about a million years to do so, and this time the mammal species were amongst the main beneficiaries, expanding in numbers and diversity and spreading into a large variety of environments across the planet. In this very real sense we humans, the descendants of these mammals, owe our existence to this catastrophic event.

The event can be dated to about 65 million years ago by radiometric techniques similar to the ones I described in Chapter 3. Rock strata that date from before this time contain fossils of the animals and plants typical of the end of the age of the dinosaurs, from the Cretaceous period. The younger strata reveal the loss of many of these species and also record the rise and radiation of the mammals during the subsequent Tertiary period. The separation between these strata, known as the K-T boundary (from the German spelling of Cretaceous-Tertiary), can be dated only to within about half a million years, and for this reason the length of time over which the extinctions actually took place is uncertain to the same extent.

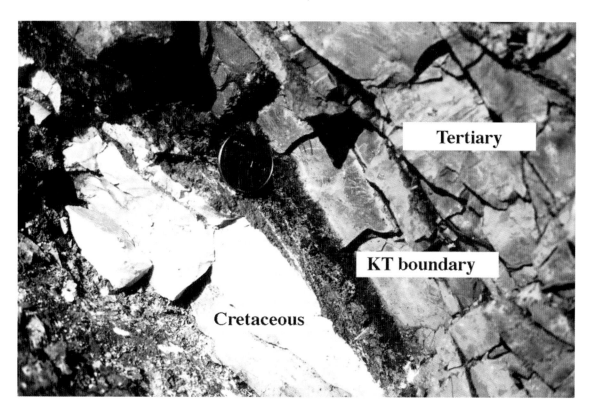

The K-T boundary is unusually distinct in the stratigraphic sequence (Fig. 7.4). In most places there is a thin seam of clay, sometimes associated with a layer of soot. However, there is also an abnormally high concentration of the rare metal iridium in the boundary layer, an imprint that seems to be world-wide. The United States geologist Walter Alvarez first identified this iridium anomaly in a rock sequence at Gubbio in Italy in the 1970s.

7.4 - The K-T boundary, a thin layer that separates rocks of the Cretaceous and Tertiary periods.

The reasons for the extinction of the dinosaurs had fascinated and absorbed both professional and amateur palaeontologists for over a hundred years. Amongst the explanations put forward had been insufficient adaptability, lack of brainpower and even an increase in the numbers of mammals that were feeding on their eggs. Alvarez, with his physicist father Luis, now proposed a completely new idea.

Iridium is a rare element at the surface of the Earth, but its concentration in the K-T boundary layer is over twenty times greater than normal. Alvarez knew that the level of iridium in meteorites is higher than in the Earth's crust and he suggested that the iridium anomaly was a result of a catastrophic collision between the Earth and a massive meteorite or asteroid. Such a collision would have vaporised the meteorite and thrown an enormous volume of material into the atmosphere, producing a cloud of dust that would have encircled the whole globe. It was even possible to estimate the size of the object. It would have been about 10 kilometres across and it would

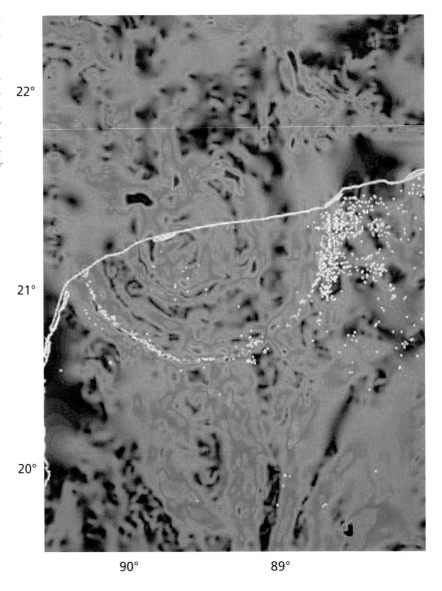

Fig. 7.5 - *A map constructed from gravity measurements obtained in the course of petroleum prospecting near the Yucatán peninsular. The Mexican coast is shown as a white line, with the Caribbean at the top. The concentric features revealed the location of the Chicxulub crater.*

have created an impact crater some 100 kilometres or more in diameter.

The theory was an attractive one, and it was supported by the discovery in the boundary layer of other signs of a cataclysmic collision. There were crystals of shattered quartz and tiny beads of glass, both characteristic of large explosions or impacts. Even so, many geologists remained unconvinced, since no crater of the right size or age was known.

Then, in 1990, the impact theory was unexpectedly and dramatically confirmed. A prospecting survey of a region near the Caribbean coast of the Yucatán peninsular in Mexico revealed the remains of an impact crater centred on a region near the small town of Chicxulub, near the city of Mérida. It had not been recognised until then because all the surface features have been eroded away or buried by sediment. The crater is 180 kilometres across and 30 kilometres deep, consistent with an asteroid of about 10 kilometres in diameter smashing into the ground at over 40,000 kilometres per hour. And, an analysis of the shattered and melted debris yielded an age of 65 million years.

An unimaginable amount of energy was released in this impact. Vaporised material, some excavated from the Earth and some from the asteroid itself, would have shot into the air to form dense dust clouds that would have spread out to cover the planet. It was this material, rich in iridium from the meteorite, that eventually settled out as sediment to form the distinctive K-T boundary clay layers.

The heat of the blast would have seared the Earth's surface over a huge area, annihilating any unsheltered surface-dwelling animals and setting fire to forests and grasslands. This vast firestorm, which could have destroyed as much as quarter of the Earth's surface vegetation, accounts for the sooty layer found at the K-T boundary.

The dust clouds would have blocked out the sunlight for many months or years, causing the loss of many plants and the demise of the animals that depended on them. The meteorite had fallen on to an area underlain by sulphur-rich gypsum deposits, releasing a huge volume of sulphur into the atmosphere and causing acid rainstorms. The effects would have been similar to the massive lava eruptions but more intense because they would have taken place over a shorter period.

These events would certainly have caused world-wide devastation, but were they sufficient to cause the end-Cretaceous mass extinctions? The asteroid theory is now widely accepted to be the most probable explanation of the K-T iridium anomaly, but there is evidence that many species were already in slow decline at the time of the impact and the full explanation, especially for the loss of the terrestrial animals, may be more complicated. For example, one of the Earth's major recurrent lava eruptions was also taking place at the end of the Cretaceous period. Over a million cubic kilometres of flood basalt were erupted in a series of sporadic outbursts lasting perhaps 500,000 years in all, forming thickly layered plateaux that have eroded into a step-like formation in western India called the Deccan Traps (*traps* is a Swedish word for stairs). In scale, this event was not far short of the Siberian eruption and it must have had similar environmental effects. It seems quite possible that the end-Cretaceous mass extinction was a result of a combination of long-term environmental, climatic and biological changes and one or both of the catastrophic events that took place at this time.

Meteors and impacts

The discovery of the Chicxulub crater prompted geologists and astronomers to debate how often massive objects have struck the Earth and estimate what effects these impacts might have had on the development of life. The Hadean was the period of extensive bombardment, but our understanding of the way planets formed, presented in Chapter 2, suggests that there is still a large amount of interplanetary debris remaining to be swept up. The impacts have certainly not ceased. The Earth continues to be bombarded with meteoric debris at the rate of 200 tonnes each day, though the vast majority of this is in the form of small particles that burn up in the atmosphere. However, US Air Force early warning satellites detect about 30 explosions each year in the upper atmosphere of a kilotonne or more, caused by disintegrating meteors of a few metres in diameter. Larger fragments fall to Earth as meteorites, sometimes causing damage to buildings or cars and occasionally even striking and injuring people or animals.

More rarely, much larger objects hit the Earth. Barringer Crater in the Arizona desert, also known as Meteor Crater, is 1.2 kilometres in diameter, the result of the impact by an object about 50 metres across that took place some 49,000 years ago. Much closer to us in time was a major impact that occurred as recently as 1908. On June 30 of that year, there was an explosion in the remote Tunguska region of central Siberia that seems to have been the result of the disintegration at a height of several kilometres of a similar-sized object entering the atmosphere. The shock-wave, which shook buildings 800 kilometres away, was picked

7.6 - The Barringer or Meteor crater in the Arizona desert.

7.7 - Aorounga crater in the
Sahara Desert, northern Chad.
The crater, which is about 200
million years old, has a diameter
of about 17 kilometres. The image
was obtained from the space
shuttle by spaceborne imaging
radar. The crater is obscured by
layers of sand, but radar can
penetrate these to reveal otherwise
invisible details of geological
structure.

up on seismographs all over the world. The blast burned and flattened trees over an area of 2,000 square kilometres, the size of a large city. Many animals were killed and a number of the nomadic people whose camps were scattered through the area were injured, some dying from shock. The energy released was equivalent to the detonation of two or three hydrogen bombs. It can be imagined what devastation would have resulted if the explosion had taken place above a city. St Petersburg, for example, is at a similar latitude and only a slightly different trajectory could have resulted in a major disaster.

7.7 - Aorounga crater in the Sahara Desert, northern Chad. The crater, which is about 200 million years old, has a diameter of about 17 kilometres. The image was obtained from the space shuttle by spaceborne imaging radar. The crater is obscured by layers of sand, but radar can penetrate these to reveal otherwise invisible details of geological structure.

There is no sign of a crater at the Tunguska site, presumably because the exploding object disintegrated in the air. However, there are about 200 known impact craters on the Earth, many of which have only recently been identified from satellite radar or photography. The 17-kilometre Aorounga crater (Fig. 7.7) in the Chad desert, one of a chain, was discovered this way; although it shows up clearly in radar surveys it is invisible from the ground because the drifting Sahara sand has filled it in. Of course, the known craters represent only a fraction of the post-Hadean impacts, since erosion and mountain-building normally erase the signs of their existence after a few hundred million years.

It is estimated that Tunguska- or Barringer-sized events happen every few hundred years whereas impacts as large as Chicxulub probably occur only about once every 100 million years. However, even the impact from a two-kilometre asteroid, though not big enough to initiate a mass extinction by itself, would cause global destruction of vegetation and widespread loss of life. If such a disaster occurred today, something like a quarter of the human population would die. It is thought that a calamity on this scale could occur about once every 300,000 years.

Do any of the other known terrestrial craters represent impacts that might be linked to mass extinction events? In April 2000 a previously unrecognised 130-kilometre crater was discovered near Woodleigh in Western Australia. Its age is estimated to be around 250 – 360 million years and there have been suggestions that it might be associated with the greatest extinction of all, at the end of the Permian, but this proposal is extremely speculative since the uncertainty in the date is very large. However, some end-Permian rocks from China and Japan do have unusually high concentrations of helium and argon isotopes, similar to carbonaceous chondrites, hinting at a possible impact at this time. But, there is no end-Permian geological seam known anywhere with an iridium anomaly or any other of the distinctive signatures of the K-T boundary layer.

Hence, the coincidence of the Chicxulub impact with the end of the Cretaceous remains the only clear-cut example so far. In fact there is a much better correlation between the dates of basalt lava eruptions and periods of species decline. Amongst the 15 or so known eruptions in the last 500 million years, perhaps six or eight could be associated with major or minor extinction events.

In any case, it shouldn't be assumed that mass extinctions always have to be linked with catastrophes like impacts or eruptions. There is a school of thought that says that from time to time evolution brings the world to a point of criticality where the Earth system becomes overstretched for one reason or another. The extinction of one species leads to the extinction of another and then another in a chain reaction that ends up in widespread collapse. In this picture, a major change in the climate or a catastrophic event serves only to trigger or intensify an extinction that is on the brink of happening anyway.

In fact many ecologists are worried that we are on the verge of a sixth great mass extinction at this very moment. The rapid growth in the numbers of human beings on the planet is already putting enormous stress on the Earth system. Early humans were probably responsible for decimating or exterminating many large mammal species such as the mammoths in Europe and the native American horse in North America, either by hunting or by introducing disease. In many different places round the world, the disappearance of the big mammals took place at different times, but it generally coincided with the arrival of humans. Today, animals are still hunted for food, medicine and fur, or because they are seen as a threat to life or crops. At the same time, there is relentless destruction of natural habitats. On top of this, climate change induced by global warming is a threat to the whole planet. It is estimated that the present rate of species extinction through loss of tropical forests is already some hundreds or even thousands of times higher than normal.

A spur to diversity

The remaining chapters go on to consider the possible location and nature of life outside the Earth. Throughout these chapters we will need to bear in mind what we have learned about the essential nature and needs of terrestrial life, and in particular its evolution, which has both influenced and been influenced by the planet. At the same time we must remember that the biological richness and diversity that we observe around us is very definitely the outcome of the particular sequence of adventures that the Earth has been through, including a series of catastrophic and unpredictable accidents.

The normally leisurely pace of the evolution of life on the Earth is intimately bound up with and largely driven by slow changes in the climate and environment. These changes are propelled by the internal geophysical forces that continually shape and reshape the planet and moderated by an important extraterrestrial influence, the fluctuating amount of energy received from the Sun.

The outcome has been an increase in the diversity and complexity of life. New species appeared while others become extinct. More complicated organisms developed alongside longer-established simpler life forms and the total numbers of species grew steadily.

However, this growth was intermittently interrupted by intervals of species loss, sometimes of wholesale mass extinction. The reasons for the mass extinctions are not obvious but they may be associated with periods of major climate change and perhaps a process of natural self-limitation. They are almost certainly on occasions accelerated by cataclysmic events such as meteoric impacts and volcanic flood basalt eruptions.

Whatever the cause, each mass extinction resulted in the global catastrophic collapse of the environment, with recovery typically taking millions of years. Moreover, each such crisis was followed by a spurt of new biodiversity and the emergence of a range of different types of life. From time to time, the board gets swept clean for a fresh start. When will the next time be?

Part III:
Life and the Universe

8 Looking for life in the Solar System

Early in June 2003, a Soyuz Fregat rocket will lift off from the Baikonur Cosmodrome in Kazakhstan carrying a European Space Agency mission called Mars Express. The journey to Mars will take six months, at the end of which the spacecraft will be put into orbit around the planet. However, just before that happens, five days out from its destination, a small package will detach itself from the craft and head directly for the surface of the planet itself. A heat shield will protect it from temperatures of up to 4,000°C during its passage through the Martian atmosphere, then a parachute will slow its descent and three man-sized gas bags will inflate to protect it as it bounces hundreds of metres across the barren landscape. Finally, the gas bags will separate and roll away, dropping a strange-looking object onto the surface; this will be a silvery disc just 60 centimetres across. The hunt for life on Mars will have begun.

8.1 - Model of the Beagle 2 Mars lander with its solar panels deployed.

This unusual landing craft, designed by a consortium of British universities and space engineering companies, is called Beagle 2. The name celebrates the *Beagle*, the ship on which Charles Darwin made the epic voyage that led to his writing of *On the Origin of Species*. When it comes to rest, the lander will first unfold itself, laying out a series of solar panels to provide a source of power (Fig. 8.1), and then set a number of experiments to work.

There will be cameras to survey the scene and relay pictures of the Martian desert back to Earth. However, the overall aim of the on-board experiments is to look for the signs of life. These will sample the atmosphere, the moisture content and the chemical composition of the rocks. There is no real expectation of finding living organisms at the surface, but the existence of carbonate minerals or organic residues would be good indicators of the presence of life in times gone past.

The most curious experiment of all has the official name of PLUTO, the Planetary Undersurface Tool, though it is more colloquially known as 'the mole'. This is a little device that works its way across the ground in a series of small jumps, propelled by an internal compressed spring. It can travel about five metres from the lander on the end of its cable and it can even burrow, hammering itself into the ground or under boulders using the same mechanism. Rock and soil samples are collected in its jaws and then recovered for analysis at the lander by winding the device in.

Beagle 2 will be very rapidly followed by other spacecraft. A number of planned NASA Mars landers will be equipped with roving vehicles that will carry out surface measurements and search for water. The main objective of these missions is to find any evidence that life once existed on Mars. Why Mars? Of all the places in the Universe, why should Mars be selected for all this interest and attention?

Searching for life

The search for life beyond the Earth using space probes is restricted for now and for the foreseeable future to the investigation of the Solar System. We simply do not possess the technology to explore at any greater distance because the stars are far too remote and inaccessible. Within the Solar System, a spacecraft takes many months or years to travel between the planets, but it would take tens of thousands of years with existing propulsion systems to reach even the nearest star. Mars, of course, is one of the nearest planets to the Earth, and therefore one of the most accessible, but this is not the main reason for turning our attention to it in the first instance.

The Earth has achieved a state of long-term stability that is not only conducive to life but also, in a very real sense, maintained by life. Plant life extracts carbon dioxide from the atmosphere and helps keep the Earth cool by reducing the greenhouse effect. In the process it keeps the level of oxygen in the atmosphere high. As a result of this temperature regulation, the Earth has been able to retain its oceans for thousands of millions of years.

There are no other worlds like this in the Solar System. They are either rocky or icy worlds or else gas giants, and none of them now possesses surface oceans of water, nor atmospheres rich in free oxygen. But is it possible that any of them might have had oceans in the past, or might have retained hidden reservoirs of water to this day? If so, these would be the best places to search for the signs of life, past or present.

The most realistic place to start this quest is in the so-called habitable zone, within which the Sun heats the surfaces of planetary bodies to the temperature of liquid water. The Earth and Moon orbit comfortably in the middle of this zone, Venus lies near its inner, hotter boundary and Mars lies near the colder outer edge.

We already know enough about the Moon to be able to rule it out as a hospitable place for life. Its geological history is known from the rocks brought back to Earth by the astronauts of the Apollo programme. Its ancient surface has scarcely changed in the last 3,000 million years; unlike the Earth, the Moon has been geologically inactive during this time. The youngest areas are the dark plains called the *maria* which were formed from volcanic outpourings of between 3,000 and 3,700 million years ago. The highlands, over 3,900 million years old, are densely covered in craters, a result of the same bombardment that the Earth suffered in the Hadean era.

There is no atmosphere, since the Moon is too small and has insufficient gravity, only one sixth of the Earth's, to retain one. Without this insulating layer, the temperatures veer from −110°C during the lunar night to 130°C when the Sun is overhead and the surface is unprotected from solar ultraviolet radiation. Under these conditions, there could be no surface water. It would evaporate during the heat of the lunar day and rapidly be lost into space. Indeed there is no evidence that the Moon ever possessed any significant quantities of water, although it is possible that a certain amount of ice remains in permanently shaded regions where the Sun cannot melt it, such as in large craters at the poles. None of the Moon samples show any traces of life and, in short, all the evidence points to the fact that it would be fruitless to search for any.

If anything, the planet Venus is an even less promising prospect. Venus is almost a twin to the Earth in size and mass. Yet today it has a very dense atmosphere, largely composed of carbon dioxide with smaller amounts of nitrogen and sulphur dioxide and only a little water vapour. Why should it have developed so differently from the Earth?

Like the early Earth, Venus may originally have had oceans, but because it is closer to the Sun, the surface temperatures would have been higher and the evaporation rates would have been greater. Both water vapour and carbon dioxide are greenhouse gases and it seems that Venus experienced an extreme runaway greenhouse effect. This resulted in the complete evaporation of the surface water and its eventual loss from the atmosphere into space. Some see Venus as a possible model for the Earth should a runaway greenhouse effect ever take hold here. The average surface temperature on Venus is now a blistering 480°C, well above the melting point of lead, and the atmospheric pressure is 90 times that of Earth. A number of Soviet landers in the Venera series touched down on the surface in the 1970s and 1980s but none of them survived more than an hour in these merciless conditions.

The surface of Venus is invisible from above its atmosphere because it is completely covered by layers of clouds, which are composed of sulphuric acid rather than water vapour. Despite this, comprehensive maps of the planet have been made from orbiting satellites, using microwave radar that is able to penetrate the clouds. NASA's Magellan spacecraft, which orbited Venus and surveyed it in 1990-93, was able to see details on the surface less than a kilometre across. Much of the terrain is old, cratered and highly deformed but there are also extensive lava flows and many volcanoes. A number of these appear to have been recently active – for example, the flows are free of impact craters, and this is a sign of a relatively newly-formed surface. It is quite possible that Venus remains volcanically active today, although this has not yet been established. In view of the extreme conditions, particularly the high temperatures and the absence of water, Venus is likely to be a difficult and unrewarding place to search for life.

Mars

Mars, the only other planet in the habitable zone, is near its outer, colder edge, farther from the Sun than the Earth. Mars is smaller than the Earth, with about half its diameter and only about a tenth of its mass. In September 1997 NASA put a spacecraft called Mars Global Surveyor into orbit around the planet which since March 1999 has been returning detailed maps and pictures of the surface. It seems that Mars has two quite distinct hemispheres. The southern hemisphere is rough and heavily cratered, corresponding to the bombardment era and indicating an age of 3,500 million years or more. However, most of the northern hemisphere is a low, level plain, a younger surface with few craters. Although now quiescent, Mars has certainly been geologically active since it has a number of dormant or extinct volcanic regions, and there is evidence of crustal movement in the past.

Today there is only a thin carbon dioxide atmosphere that provides very little insulation and as a consequence the surface temperatures are low, averaging –50°C and almost always remaining below zero even at the equator. There is a little water in the atmosphere in the form of vapour and ice in very sparse clouds, but no liquid water at the surface. The shallow Martian icecaps, a mixture of carbon dioxide ice and water ice, wax and wane with the seasons. The thin atmosphere affords no protection from solar ultraviolet radiation.

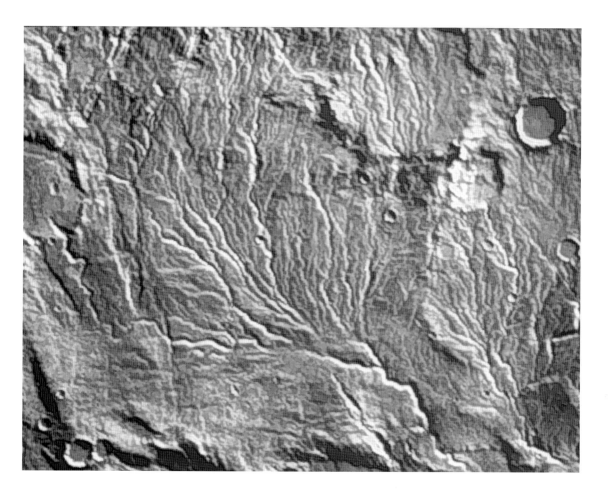

What has excited astronomers and biologists alike is the presence of a number of surface features that look very much as if they have been formed by running water. Images from NASA's Mariner and Viking missions in the 1970s showed localised but extensive systems of what are apparently dry river valleys, with branching channels and islands (Fig. 8.3). There are also many smaller gullies on the walls of craters, characteristic of the runoff of water.

8.3 - A Martian dry river network in the southern highlands, seen by NASA's Viking orbiter. The view is about 160 kilometres across.

Whereas some features must have formed slowly as part of a long-lasting drainage system, there are larger valley networks that look as if they were formed in catastrophic floods. They do not originate in drained lakes but from regions of chaotic debris, where great quantities of subsurface water appear to have burst out of the ground. These may have been released by the heat from volcanic or other geological activity or even by a meteor impact. Most of the valley networks are in ancient terrain and so must be older than about 3,500 million years, though there is evidence that flood events have occurred throughout Martian history. It is likely that substantial amounts of water remain beneath the surface to this day, frozen into the permafrost and possibly even kept liquid in subterranean reservoirs by residual geothermal heating.

As well as these signs of both lasting and episodic flows of water across the surface in the past, there is evidence that an ancient ocean may at some time have covered the whole of the northern hemisphere. The Mars Global Surveyor showed that the northern plains are on average five kilometres lower than the southern highlands, with a boundary having a narrow range of elevations that could once have been a shoreline. In addition, some of the large outflow channels run into the plains, looking like dried-up rivers that used to run into a sea.

The presence of surface water in the past implies a warmer, thicker atmosphere than at present. The atmosphere of Mars was probably quite substantial for the first 1,000 million years of its history, with warmer and wetter conditions prevailing for several hundred million years. The early Mars could well have been quite like the early Earth.

What happened to its atmosphere? On Earth, the carbon cycle sucks the CO_2 from the atmosphere and stores it underground as carbonates from where it is recycled by geological processes. Perhaps this also happened on Mars at first, but the geological cycling switched off or slowed very early in its history, the recycling stopped and the atmosphere was no longer replenished by volcanic gases. Because of the planet's smaller mass and lower gravitational attraction, the atmosphere was more easily lost to space. As it thinned, the oceans evaporated or ran underground and Mars cooled to the state in which we see it today.

8.4 - A high-resolution image from the Mars Global Surveyor shows a large region of layered sedimentary rock in south-western Candor Chasma on Mars. The picture covers an area 1.5 kilometres by 2.9 kilometres.

109 yd
100 m
1.5 aspect ratio

It is the existence of water on Mars that makes it an enticing place to start the search for extraterrestrial life. Life appeared on Earth sometime before about 3,500 million years ago, at a time when the conditions on Mars and Earth could have been very similar. So, why not on Mars too? If life did ever develop there, signs of it should remain. Images from the Mars Global Surveyor have picked out sites where layers of rock can be seen, formed perhaps as much as 4,000 million years ago by sedimentary deposits in huge lakes; these represent further proof of the previous existence of water on the planet (Fig. 8.4). These are exactly the kinds of rocks in which fossils are preserved and these sites would be a top priority target for future space missions. The discovery of Martian fossils of this age would be doubly important since there are no fossils this old on Earth. At the same time as establishing the existence of life on Mars, they would give valuable information about the earliest stages in the evolution of life, from a time before the terrestrial record begins.

8.5 - The view from the Martian surface as seen by the Viking-2 lander at Utopia Planitia.

EARTH, LIFE AND THE UNIVERSE

Could living organisms remain on Mars today? Back in 1976, NASA landed two Viking spacecraft onto Mars with the main aim of investigating the surface environment of the planet. These missions were highly successful, and over the next few years they obtained a wealth of superb meteorological, atmospheric and geological data (Fig. 8.5). As part of the package, three experiments were included that were designed to stimulate and detect any biological activity that might be occurring in the Martian soil. The results of these experiments were somewhat inconclusive because of the difficulty in distinguishing biological from inorganic chemical reactions but it was generally agreed that they gave no positive indication of living organisms at either site.

In fact the expectations of finding existing living organisms on the surface of Mars are not high, given the sub-zero conditions and the high ultraviolet flux. However, our knowledge of the astonishing adaptability of life on

Earth gives some slight hope that Martian life may have managed to survive or remain dormant in more sheltered locations. Given that bacteria and lichen can be found thriving inside Antarctic rocks, it is just about conceivable that the same could have happened on Mars, though with temperatures even lower than in Antarctica, this seems unlikely. Perhaps dormant organisms remain in the Martian permafrost or deep below ground in subterranean reservoirs. To investigate these ideas will require sophisticated space missions of a kind that are only now reaching the early planning stages.

Rocks from Mars

The negative results from the Viking lander biological experiments were followed by something of a slackening of interest in continuing to search for life on Mars. At the same time, other Viking measurements enabled geologists to prove that certain rare types of meteorites found on Earth are actually fragments of Martian volcanic rocks. The gases trapped in these meteorites were found to match the composition of the Martian atmosphere determined by the Viking instruments. In other words, we already had rocks from Mars here on Earth.

Only a dozen or so Martian meteorites have been discovered so far. Radioactive dating shows that most of them are quite young, having formed between 200 and 1,300 million years ago, and so they must have come from the younger parts of the Martian surface. How did they get here? Just like the Earth, Mars has from time to time sustained massively large meteoric impacts. It seems that some of these impacts were sufficiently powerful that as well as forming a crater, some of the resulting debris was thrown upwards so violently that it escaped from the surface altogether. This would have been relatively easy because of the low Martian gravity and thin atmosphere. Having escaped from Mars, it would be just a matter of chance whether a fragment ever reached the Earth.

Interest in the question of life on Mars was suddenly rekindled with a vengeance in August 1996 when, at an historic press conference, a NASA team led by David McKay announced that possible fossil life had been discovered in one of the Martian meteorites named ALH84001 (Fig. 8.6). The media coverage and the prolonged scientific argument that followed led to a great intensification of research into the whole subject and the initiation of a 20-year NASA space programme called *Origins* to search for life-sustaining planets within and beyond the Solar System. The discoveries announced at the press conference, however, have remained controversial.

ALH84001 is much older than the other Martian meteorites, with an age of around 4,500 million years, and it must have been blasted from the ancient crustal regions. Weighing nearly two kilograms, it was collected from an Antarctic ice field known as the Allan Hills in 1984 (hence the ALH84 part of its name). Part of the reason for all the excitement was the discovery within the meteorite of unusual organic compounds and mineral features that are

8.6 - The Martian meteorite ALH84001. The scale is shown by the one centimetre cube at the bottom right.

characteristic of biological activity. All living things leave this kind of evidence of their existence behind even where no visible fossil form can be seen.

The NASA group also discovered long chains of magnetic crystals that they claim could have come only from living organisms. On Earth, similar chains are formed by bacteria, which use them as miniature compasses. About a thousand chains have been found in ALH84001, with the uniform shapes and sizes that are typical of a biological origin.

However there are alternative non-biological explanations for most of the observed features. Moreover, there is the further possibility that the sample was contaminated by terrestrial organisms after the meteorite arrived on the Earth – it is thought to have fallen about 13,000 years ago – though the scenarios put forward for the way in which this might have happened also have their difficulties.

8.7 - Bacteria-shaped objects found in ALH84001.

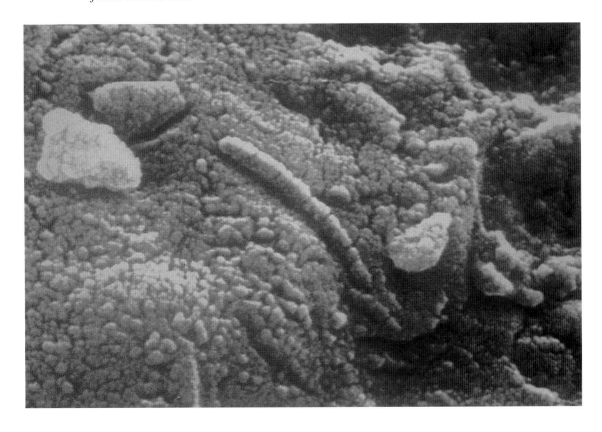

EARTH, LIFE AND THE UNIVERSE

But the most dramatic signs of Martian life were minute structures that enthusiasts interpreted as microfossils having the appearance of Earth bacteria (Fig. 8.7). The largest of these, christened BSOs or bacteria-shaped objects, are about a tenth of a micron in length (recall that a micron is 1/1000th of a millimetre), though most are five or ten times smaller.

For most microbiologists, the suggestion that these were fossils of Martian organisms was very hard to accept. The problem is that the BSOs are far smaller than Earth bacteria; in fact the smaller ones are not much thicker than a typical cell wall. This is so small that it is hard to see how they could contain the complex biological molecules like the nucleic acids and the proteins that an organism needs to function.

Then only a couple of years later, the argument was given a new twist by the discovery of similarly tiny structures, given the name nanobacteria or nanobes, in terrestrial rocks. Scientists from the University of Queensland discovered the nanobes in Australian sandstone several kilometres below the seabed. These are composed mainly of carbon, nitrogen and oxygen and appear to contain membranes and a cell wall. There are also indications that they contain DNA.

Some microbiologists have suggested that nanobes are a form of cellular debris, or fragments of larger living cells. In this case, the BSOs in ALH 84001 might not be complete fossil bacteria but could perhaps be fossils of bacterial fragments. Others remain sceptical, believing that they are simply non-biological artefacts. Few scientists unreservedly accept that the BSOs are fossil Martian organisms and no agreement has emerged on whether or not the meteorite ever contained Martian life. In short, the evidence is inconclusive, and neither side in the debate is able to prove its arguments convincingly.

The only way to resolve the problem is to explore Mars itself and to find more conclusive evidence in the form of fossils, organic matter or, just possibly, surviving or dormant living organisms. The best places to visit are the valleys, sedimentary strata and other landscape features that look as if they were formed by flowing water. We may have the answer within a matter of years. Beagle 2 will start the search in earnest in 2003, and the European Space Agency is considering a mission to collect a Martian soil sample and return it to Earth by around 2010. NASA is also looking at a sample return mission that could be launched in 2014, with a second in 2016. As soon as the middle of the next decade, possibly earlier, we could have the first unambiguous proof of extraterrestrial life in rocks brought back to the Earth from the Martian surface.

The ALH84001 controversy has shown how difficult it might be to recognise and agree on the evidence for any newly discovered life forms. Microbiologists regard the BSOs with suspicion because they seem to be too small to allow vital cellular functions to operate in the way we have come to understand. However, is this a fair objection? We cannot be sure whether terrestrial bacteria developed from smaller pre-bacterial organisms or, for the sake of argument, from larger cells with a simpler chemistry. In fact, if we did discover organisms on Mars, it should not surprise us at all if they were different from what we find on Earth. The whole debate is in fact a good rehearsal for the sorts of problems that are likely to be faced in the search for life in the Solar System.

The undisputed fact that meteorites from Mars have landed on Earth raises the fascinating possibility that they may have carried viable life forms with them. And perhaps the reverse journey is also possible. So if life is discovered on Mars, it may be related to Earth life; in which case the intriguing question will be, which came first?

Meteorites or comets striking the Earth seem to have played an important role at various stages of the development of terrestrial life. They probably added water to the early oceans and may originally have seeded the surface with complex organic material. Larger impacts were the cause of occasional drastic environmental changes and probably contributed to one or more of the mass extinctions. And now it seems they could even have been responsible for carrying life from one planet to another.

In summary, some 4,000 million years ago, conditions on all three habitable zone planets, Venus, Earth and Mars, may well have been quite similar, and at that time life could possibly have existed on all three of them. If it ever did develop on Venus, it seems unlikely that it has survived. But, the evident signs that water was abundant on Mars in the past and the possibility that it remains in underground reservoirs to this day makes this the most attractive target for the first serious search for extraterrestrial life.

Beyond the habitable zone

At first sight, the chances of finding life within the Solar System outside the habitable zone seem slim. The only planet closer to the Sun than Venus is Mercury, a bleak, airless and waterless world where temperatures cycle between 470°C in the daytime and −180°C at night. Farther out from the Sun than Mars, the gas giants have no solid surface, or rather the small solid cores that they do possess are buried deep under massive atmospheres at intolerably high temperatures and pressures.

This leaves only the smaller bodies of the Solar System. As explained in Chapter 2, some of these, especially the giant planet satellites, are quite large. Figure 2.3 shows that Ganymede and Callisto, the largest satellites of Jupiter and Titan, the largest satellite of Saturn, are comparable in size to the planet Mercury. Even Io and Europa, the next largest satellites of Jupiter, are similar in size to the Moon. All of them lie beyond the habitable zone and so they might be expected to be icy, permanently frozen worlds. In fact the story is much more complicated and more exciting than this.

The four largest satellites of Jupiter are collectively known as the Galilean satellites after their discoverer. Not only are they large, but they orbit comparatively close to the giant planet and as a result they experience the effects of its intense gravitational field. Each satellite suffers a stretching force, since the attraction on the side of the satellite that faces Jupiter is stronger than that on its far side. A purely liquid body would elongate in the direction of attraction; this is the cause of the ocean tides with which we are familiar on Earth, raised by the gravitational attraction of the Moon. But, a solid body also flexes and distorts, producing friction and heat just as a rubber ball would if it were repeatedly squeezed. The effect generates a significant amount of heat in the interiors of the Galilean satellites. It is strongest on Io, the closest to Jupiter, and is progressively less on Europa, Ganymede and Callisto.

In the case of Io, intense internal tidal heating keeps the satellite in a permanent state of active volcanism. Pictures from the Voyager spacecraft and the Galileo mission show that the surface is entirely covered in volcanic vents and lava flows, with active eruptions constantly in progress. Io is easily the most volcanic body in the Solar System.

8.8 - This enhanced-colour picture from the Galileo spacecraft shows a 60 kilometre-wide region of the icy surface of Europa known as Conamara Chaos. Ice floes can be seen that have apparently shifted to new positions before water from below welled up and froze.

In direct contrast, Europa, a little farther out from Jupiter than Io, has one of the brightest surfaces in the Solar System and for some time it had been suspected that it was covered by an icy crust. This was confirmed by images from the Voyager flyby missions of the 1970s, but it was not until the late 1990s that the Jupiter orbiter named

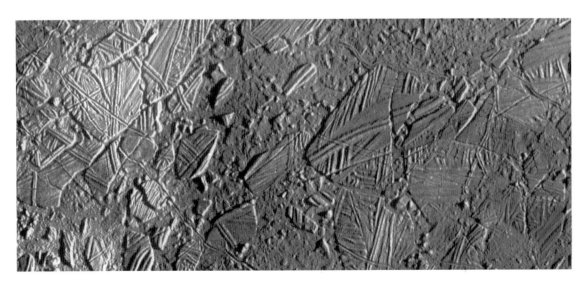

Galileo started to send detailed pictures of the Europan surface back to Earth. These images revealed a totally unexpected panorama that amazed and delighted planetary scientists and astrobiologists alike.

Instead of a heavily cratered landscape, they showed an icy surface criss-crossed by cracks, ridges and grooves (Fig. 8.8). There are crustal plates resembling polar ice floes and pack ice on Earth. In some places the surface seems to have shifted and been broken apart, with the gaps filled from below with fresh melted material. There are few impact craters and the whole impression is of a young, recently refreshed surface.

The overall density of Europa is quite high, which shows that it is very largely made of rock, probably with an iron-rich core, and the icy crust is at most 200 kilometres thick. There is significant internal tidal heating because of Europa's proximity to Jupiter, though not nearly as much as on Io. What all these observations hinted at, and what excited the scientists, was the possibility of the existence of a large ocean under the ice.

Could the internal heating be generating enough energy in the rocky core to keep an ocean liquid under the icy surface? If so, this could explain the appearance of many of the surface features. We could be looking at an icy crust floating on liquid water. Perhaps the surface has been formed and reformed in a similar way to the oceanic and continental crusts on Earth, powered by convected heat from the core below. Alternatively, perhaps the ice cracked as it flexed under the rise and fall of the tides raised by Jupiter, allowing water to flow up to the surface and freeze. Today's tides would drive a floating ice crust up and down by some 30 metres in each cycle, not enough to explain the larger fractures and ridges that are observed, but the effect may well have been much greater in the past. The circumstantial evidence for oceans is impressive, but how could their existence be proved?

The most persuasive evidence unearthed so far also comes from the Galileo spacecraft. Galileo detected changes in Europa's magnetic field that seem to arise in weak electric currents in a conductive layer near the surface of the satellite. Salty water is a conductor of electricity, though ice is not, and so the most plausible explanation is that the magnetic fields are being generated within a watery or slushy layer beneath the crust. The observations are consistent with the presence of a briny ocean up to 60 kilometres thick, starting somewhere between five and ten kilometres below the surface.

If such an ocean exists, could it support life? With liquid water and an abundant supply of chemicals, there seems to be no compelling reason why it could not. It might now be life in a dark, cold, salty environment, but that would be no worse than in many of the harsher places where it has survived on Earth. Moreover, it may not be so cold. If the rocky core is hot enough, there could be hydrothermal systems on the ocean floor that could sustain life forms like our own deep-sea vent creatures. There is even some evidence for local hotspots like this; there are frozen ponds on the surface that look as if they could have been caused by localised convection heating from below.

There is even a slight possibility that life has percolated to the surface and colonised the fissures in the ice. The periodic tidal stresses may repeatedly open and close the cracks on the Europan surface, allowing water to rise up from below and form a freezing layer as it reaches the surface. The water could perhaps carry living organisms with it and in regions like this the ice may be thin enough for sunlight to reach through and drive photosynthesis. Certainly the fissures imaged by Galileo show subtle colouring that could be deposits of salts or organic molecules.

The question is unlikely to be settled until further generations of space probes are designed and launched. NASA plans to send a mission to Europa that will enter orbit around the satellite in 2009. It will be equipped with ice-penetrating radar to confirm the existence of an ocean and survey the depth of the ice. If successful, it is hoped that it will be followed by a lander capable of drilling through the ice and releasing a submarine explorer. This is clearly a technically ambitious project that will need very careful planning.

Ganymede and Callisto, the two other large satellites of Jupiter, are also rocky-icy worlds. However, neither of them experiences as much tidal heating as Europa. Ganymede, the largest satellite in the Solar System, has a rocky core but a much thicker icy crust than Europa, whereas Callisto seems to consist of a largely undifferentiated rocky-icy mixture. Magnetic measurements hint that both might contain watery layers, though the evidence is less clear than on Europa and the layers are likely to be thin and deep-lying.

With Europa and possibly Ganymede and Callisto, the very short inventory of the potential locations of liquid water in the Solar System is complete. However, there is one other curious world we should mention before moving on. Titan, the largest satellite of Saturn, is unique amongst the satellites in that it is blanketed by a substantial atmosphere. This is nearly all composed of nitrogen, obscured by a haze of droplets of hydrocarbons (which are compounds containing only carbon and hydrogen), for example methane and ethane. These hydrocarbons and other compounds formed in the atmosphere probably fall as 'rain' or 'snow' onto the icy bedrock.

Although water is present at the surface, it must all be in the form of ice, since the temperature is −180°C. Nevertheless, a methane-ethane mixture could remain a liquid and could form lakes or oceans. Is it possible that cold organic processes could result in some exotic kind of living organism under these conditions? In 2004 the European Space Agency's Huygens probe will arrive at Saturn after a journey lasting seven years aboard NASA's Cassini spacecraft. Huygens will descend by parachute through Titan's atmosphere and onto the surface, but until it arrives, it is not known whether it will land on layers of tar-like hydrocarbon solids or splash down into a methane-ethane sea.

In conclusion, the most promising extraterrestrial sites to search for life within the Solar System, existent or extinct, are the planet Mars and the satellite Europa. What might we realistically expect to find? In either case, the chance of encountering advanced life forms is small. There are hopes of finding fossil evidence on Mars left over from a more congenial period of its history, and the outside possibility of coming across surviving or dormant micro-organisms in underground refuges. And, life may continue to thrive in the warm, extensive oceans of Europa. If present on either of these, it is likely to be discovered by one of a number of space probes planned within the first two decades of the twenty-first century.

9 Looking for life in the Universe

If our understanding of the formation of our own Sun and Solar System (described in Chapter 2) is correct and if this is part of a universal phenomenon, a by-product of the cosmic cycle, then we might expect the same process to have occurred repeatedly throughout the Universe.

It is estimated that there are 100,000 million stars in our galaxy alone. Of course, not all of them are like the Sun. Some are much more massive, and they exhaust their nuclear fuel and expire after only a few million years, not long enough for life to emerge on any planets that may have formed around them. Other stars are much less massive and less luminous than the Sun. Their habitable zones are small; for a star with one hundredth of the Sun's luminosity it would be ten times closer to its parent star than it is in the Solar System. A planet in this zone would have such a tight orbit that it would become locked into slow rotation, with one side perpetually facing the star and the other facing away. One hemisphere would boil, the other would freeze and no oceans could form in either.

So it is most likely that if any Earth-like planets existed, they would be in orbit round solar-sized stars. However, something like a third of all stars are between a half and one-and-a-half times the mass of the Sun, sufficiently similar that it is reasonable to suppose that systems of extrasolar planets and satellites like our own could be common in the Universe. Moreover, if the processes that led to the formation of the Earth are truly universal then there should be many Earth-like planets orbiting other stars, and perhaps other kinds of worlds also capable of supporting life.

How could this idea be confirmed or disproved? Until very recently our theories were based on just one planetary system, our own Solar System. Despite strong suspicions that extrasolar systems must exist, there was no evidence for any others, let alone how similar they might be to ours.

The problem is that the direct observation of extrasolar planets is at present impossible. We cannot simply turn our telescopes on other stars and observe planets in orbit around them. The distances to the stars are vast. Even the nearest star, Proxima Centauri, is at a distance of 4.2 light-years; that is, its light takes 4.2 years to reach us. Any accompanying planets would be so faint and be orbiting so close to the parent star that they would be lost in the glare of the star itself.

Size of Pluto's Orbit

9.1 - *This Hubble Space Telescope infrared image of the region near the star β Pictoris shows a dark region in the centre where the light from the star itself has been deliberately masked off. The bright extensions on either side are the outer parts of a disc seen nearly edge-on.*

Nevertheless, distinct signs of planetary formation can be seen in infrared images of the regions surrounding one or two of the nearest stars, such as β Pictoris (Fig. 9.1), which reveal disc-shaped clouds of dust and gas. Using the infrared part of the spectrum makes it easier to distinguish the disc from the star because these discs emit most of their radiation as heat, whereas the star emits most of its radiation as light. From the nature of the emission, it is possible to infer that the discs are probably made of icy, carbon-rich and silicate particles in orbit around the star, the sorts of materials from which planets are expected to develop. Some discs even show gaps at their centres of about the right size for a planetary orbit, where it is thought that planets may already have formed.

But the most convincing evidence for the existence of extrasolar planets comes indirectly, from the effects that the planets have on their parent stars.

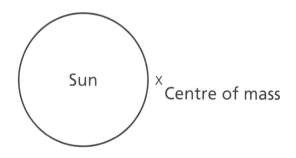

The typical representation of a planet orbiting a star would depict the planet in motion and the star occupying a fixed position in space. In this picture the planet is held in its orbit by the gravitational attraction of the star, since the mass of the star is so much greater than the mass of the planet. However, in reality the planet also has a small effect on the star. As the planet orbits, it perturbs the star's position a little so that the star also executes a miniature orbit. In fact it would be more correct to say that the planet and the star orbit around their common centre of mass, since each pulls gravitationally on the other (Fig. 9.2). For most planets the effect on the star is minute. Even in the case of Jupiter, the largest planet of the Solar System, the size of the 'orbit' of the Sun is only a little greater than the radius of the Sun itself, so both Jupiter and the Sun orbit around a point that lies just outside the solar surface.

9.2 - A star and a massive planet orbiting around their common centre of mass, which in this case is just outside the surface of the star.

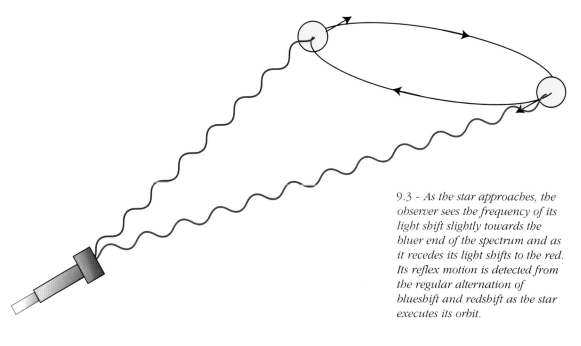

9.3 - As the star approaches, the observer sees the frequency of its light shift slightly towards the bluer end of the spectrum and as it recedes its light shifts to the red. Its reflex motion is detected from the regular alternation of blueshift and redshift as the star executes its orbit.

This motion of a star is called its reflex motion, and it was the detection of small reflex motions in a number of nearby stars that provided the first real proof of the existence of extrasolar planets. In effect, certain stars were found to be wobbling periodically back and forth in tiny orbits that pointed to the presence of invisible planetary companions. These stars were not seen literally moving from side to side in the sky; their orbits are too small for this kind of motion to be observed with present-day techniques. What was actually measured was the regular oscillation of the star's speed, alternately of recession away from us and approach towards us, as it carried round its orbit. This kind of motion along the line of sight produces a recognisable change – called the Doppler shift – in the frequency of the light coming from the star, which fluctuates periodically (Fig. 9.3).

Astronomers had long used this technique to observe certain star systems called binary stars. These are pairs of stars that orbit around each other, bound by their mutual gravitational attraction. In this case the reflex motion is usually quite large and easy to measure. Careful measurement of the amount and period of the variation in the Doppler shift allows the calculation of the masses of the stars and the distance between them. However, in the case of the reflex motion caused by a planet, the fluctuation is extremely small; so small, in fact, that until a few years ago it had been below the capability of any available equipment.

Several groups of astronomers throughout the world had been carrying out searches for these signs of extrasolar planets for over a decade. Then, in October 1995, Michel Mayor and Didier Queloz from the Geneva Observatory announced the discovery for which everybody had been waiting. For 18 months they had been studying a star about 50 light-years away called 51 Pegasi, which is very similar to the Sun. They had analysed their data extremely carefully, only releasing their conclusions when they were very confident of their interpretation. They had detected a minute amount of reflex motion in 51 Pegasi that was of a size and period that could be caused only by a planetary companion.

The Swiss group's caution was understandable, for the planet they had discovered was quite unlike anything in our Solar System. It is a giant, with a mass lying between that of Jupiter and Saturn, yet it orbits 51 Pegasi once about every four days at a distance of only 7.5 million kilometres. This is eight times closer than our innermost planet Mercury is to the Sun.

The discovery was quickly checked and confirmed at the Lick Observatory in the United States by another group that had been in the hunt, which was led by Geoffrey Marcy and Paul Butler. Alerted by the important clue that the sought-after planets might have unexpected properties, Marcy and Butler were almost immediately able to detect the signs of two more of them in observations that they had already obtained of other Sun-like stars. Both are giant planets. One with a mass at least 2.4 times that of Jupiter circles 47 Ursae Majoris at a distance of about 315 million kilometres, an orbit somewhat larger than that of Mars. The other, companion to the star 70 Virginis, is truly massive, over 6.6 times more massive than Jupiter. It follows a very strange orbit, similar in size to that of Mercury but far more elliptical than any of the planets of the Solar System.

Since then, further discoveries followed rapidly one after the other. The number of confirmed planetary companions to main-sequence stars passed the 70 mark in 2001, over half of them found by Marcy and Butler. A few of the systems contained two or even three planets.

At the same time as the search was throwing up these new discoveries, a second way of identifying extrasolar planets was being developed. It happens occasionally in some systems that a planet's orbit is oriented so that it appears edge on as seen from the Earth. If this is the case then the planet periodically passes in front of the stellar disc and, if the planet is large enough, this transit can cause a minute but measurable dimming in the light from the star. This method was used successfully for the first time to detect the transit of an otherwise invisible planet in orbit around the star HD 209458. Yet again it turned out to be a giant, quite similar to the companion of 51 Pegasi. With a mass between that of Jupiter and Saturn, it orbits very close to its star once every 3.5 days. The planet's diameter, which can be calculated from the amount of the dimming that it causes as it transits, is one and a half times that of Jupiter, confirming that it must be a gas giant with an extended atmosphere.

In the early years of the search, there were a number of false alarms. Stellar reflex motion can be the result of the gravitational pull of either a binary star companion or a planetary system and it becomes possible to distinguish between the two only when the mass of the perturbing object has been calculated. On several occasions it turned out that the cause was an unseen companion star. Many of these belonged to a class of low-mass star called the brown dwarfs, stars that are so small that when they formed they never achieved temperatures high enough for normal nuclear fusion reactions to start up.

In fact these discoveries have sparked something of a debate about the exact difference between a planet and a star. The established view, set out in Chapters 1 and 2, is that stars form from large fragments of contracting gas clouds, compressed and heated by their own gravity until they reach a temperature at which the nuclear fuel ignites and the star begins to shine. A fragment needs to have a minimum mass of about 8% of the mass of the Sun for hydrogen to commence fusion. Brown dwarfs, which fall below this limit, flare only briefly, using up a little deuterium as fuel. In this sense they are failed stars.

Planets, on the other hand, are thought to form by a process of accretion from the remains of the fragment from which the star was formed; the solar nebula in the case of the Sun. However, recent discoveries of free-floating planetary-sized objects show that they may also emerge directly from the fragmentation and collapse of interstellar clouds, separately from stars.

Jupiter, the largest planet in the Solar System, is nowhere near as massive as a star. It has only 0.1% of the mass of the Sun. However, many of the new extrasolar objects being discovered occupy the intermediate territory, significantly more massive than Jupiter yet less massive than hydrogen-consuming stars. Are they brown dwarf stars or are they superplanets? There is no clear consensus about how to distinguish between them. However, there is reason to think that no brown dwarf can exist with a mass less than 13 times that of Jupiter, and so this is an appropriate figure at which to draw the dividing line. Anything below this limit can be safely classified as a planet.

9.4 - The extrasolar planets discovered so far are all giants, yet their orbits are generally very much smaller than those of the Solar System giants. Diagram (a) gives the range of extrasolar planet masses and (b) shows the sizes of their orbits, with the terrestrial planets for comparison.

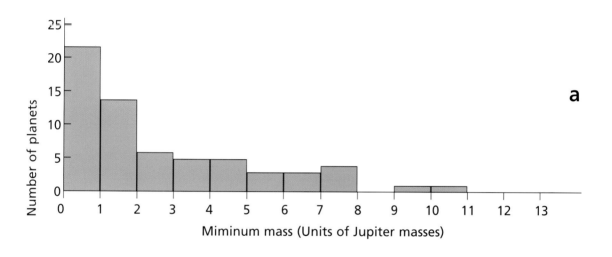

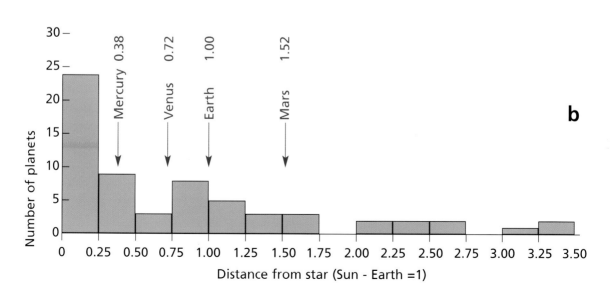

The extrasolar planets

Using this criterion, several dozen extrasolar planets are now known. The existence of extrasolar planetary systems is therefore well established and it does indeed seem that planets are common in the Universe. Yet few of the systems presently being unearthed bear much resemblance to our own Solar System.

It is hard to draw many firm conclusions from the mixed assortment of planets discovered in these searches so far (Fig. 9.4). It is evident that they are all giants, and well over half of them are significantly more massive than Jupiter. In addition, their orbits are generally very much smaller than those of the giants of the Solar System; in fact most of their orbits are smaller than the orbit of Mercury.

On the other hand, the more massive planets and the ones that orbit most closely to their stars are precisely the ones that cause the greatest reflex motion and are therefore the easiest to detect. Hence, it is likely that at the moment only this kind of planet is being discovered simply because the equipment in use is not yet sensitive enough to locate any other sort. In this case, the method could be selecting a peculiar and unrepresentative sample of planetary systems. Indeed, as the searches get more sensitive, progressively lower mass objects are now being found. In March 2000 an announcement was made of the discovery of the first two extrasolar planets with masses less than that of Saturn, orbiting two stars called stars HD 16141 and HD 46375.

The sensitivity of the detection method is only now approaching the point where Jupiter-sized planets in Jupiter-sized orbits might be detected. Because of this selection effect it is still too early to say whether high-mass close-orbit planets are the commonest kind or whether arrangements like the Solar System are more typical. On the basis of the extrasolar systems uncovered so far, astronomers have detected planets around about 5% of the stars they have observed, though this must be an underestimate of the true fraction. Even so, systems like ours may be relatively rare.

Many of the extrasolar planets have been found to travel in highly elongated trajectories around their stars. This quite unexpected property is in sharp contrast to the situation in the Solar System, where the planetary orbits differ only slightly from circles. Of the nine major planets, only Mercury and Pluto have significantly elliptical orbits yet the majority of the known extrasolar planets are far more eccentric than these. If the planet that orbits the star ε Eridani were in our Solar System, it would in turn sweep out beyond Jupiter at the furthest extent of its orbit then loop back in to pass Mars at its closest. A massive planet in an orbit like this would be very disruptive, effectively ousting any other planets from its neighbourhood and possibly eliminating Earth like planets altogether.

So these new discoveries throw some doubt on the present theory of planetary formation, which seemed to explain the origin of our own Solar System so well. The theory set out in Chapter 2 supposes that giant planets of about one Jupiter mass or less form at large distances from their star and settle into near-circular orbits.

It now seems that this picture is inadequate and needs modification. Recent developments of the theory seem to show that giant planets probably form more readily and rapidly than previously thought, and not necessarily at very great distances from their stars. Super Jupiters may well be common. In multiple systems where giants have formed close together there is a competition, a sort of gravitational billiards, in which the losers may be ejected from the system altogether and even the winners may end up in highly eccentric orbits. In the case of the Solar System, Jupiter and Saturn may have shunted their smaller siblings Uranus and Neptune sufficiently far outwards that all four were able to settle into an amicable state of stability.

What implications do these discoveries have for the existence of life outside the Solar System? All the known extrasolar planets are suspected to be gas giants, and it is unlikely that any of them would be able to support life. In any case, they nearly all orbit far too close to their stars and their temperatures must be unendurably high. There are however a very few of them that orbit within the more comfortable region of their star's habitable zone. Although the planets themselves might not be hospitable, their satellites, presuming they have any, might be more conducive to life.

Smaller worlds are likely to exist in these and other systems, but the detection of Earth-sized planets where we might expect conditions to be more favourable for the development of life is way beyond the capability of current instruments. It is impossible at the moment to say whether such planets are common or rare and their discovery will have to await the development of new generations of telescopes.

Future searches

This is a fast-developing area of research and it is certain that the continuing ground-based observations will add many more extrasolar planetary systems to the current list. However, the Earth's atmosphere limits the usefulness of what can be done from the ground, and the most important questions can be answered only by using powerful new space telescopes. Since interstellar travel remains a matter for science fiction, all the searches planned for the immediate future will have to be carried out by spacecraft that stay well within the Solar System.

NASA hopes to launch the first of these in 2005, a telescope called Kepler that will simultaneously monitor about 100,000 stars using the transit method to search for smaller, Earth-sized planets. The telescope needs to be space-based to attain the necessary precision and to avoid the blurring effects of the Earth's atmosphere. Only a fraction of the planetary orbits will be favourably aligned for a transit, but it is hoped that Kepler might detect as many as 600 terrestrial planets.

In the longer term both the United States and the European Space Agencies are designing more ambitious missions whose objective is to tackle the major questions: whether Earth-sized extrasolar planets exist, and whether or not they bear any signs of life. As part of its long-term *Origins* programme to search for life-sustaining planets, NASA intends to build and launch a telescope called the Terrestrial Planet Finder (TPF). The corresponding ESA proposal is called Darwin.

These missions will be equipped with telescopes powerful enough to capture the images of Earth-sized planets orbiting in extrasolar systems, isolating their light from the glare of their central stars. Astronomers will be able to identify terrestrial-type planets with atmospheres and analyse their composition.

The relative amounts of atmospheric gases like carbon dioxide, water vapour, ozone and methane would show whether a planet is capable of supporting life. The detection of large amounts of water vapour in the atmosphere of an Earth-like planet might indicate the presence of oceans like the Earth's, perhaps sustaining living things. The discovery of ozone would be a particularly strong indicator of the possibility of life, since this would indicate the presence of large amounts of free oxygen, which can be generated only biologically. Methane is another good biological marker in that it easily breaks down in the atmosphere and has to be continuously produced by methane-generating micro-organisms.

These missions will also carry out a census of nearby planets, giving us a much more accurate idea of the numbers and sizes of the smaller worlds in nearby extrasolar systems and telling us whether Earth-like planets in habitable zones are commonplace or rare. Astronomers also hope to learn more about how gaseous and rocky planets form from the discs of material that surround young stars. This would help to verify or invalidate the predictions of the present theory that smaller, rocky planets form in the hotter regions close to their stars and the gaseous giants form in colder regions farther out.

9.5 - An artist's impression of a
possible configuration for the
Darwin mission. The seven
individual telescope mirrors, each
of 1.5 metres diameter, are
separate free-flying spacecraft.
Incoming infrared radiation is
combined at the central hub.

The use of space telescopes will overcome the main obstacle presently hindering the observation of extrasolar planets, the fact that the contrast between a planet and its star is so great; the planet is too faint, the star is too bright and they are both too close together. To separate the two, it is necessary to obtain sharply detailed images that

suppress the light from the star yet remain sensitive enough to detect the planet. The combination of these requirements can be realised in a particular telescope configuration called an interferometer. The contrast can be further reduced greatly by observing the in the infrared part of the spectrum. As it happens, the infrared is also the most useful waveband in which to analyse the composition of planetary atmospheres.

These are no ordinary telescopes. Although the designs will not be finalised for some time to come, it is already clear that they are gigantic in concept. One TPF proposal is for four 3.5-metre mirrors in a free-flying formation and one possible design for Darwin is for six or seven one-metre telescopes spread over a baseline many hundreds of metres across, combining their light at a central station (Fig. 9.5). Each of the mirrors will have to be manoeuvred and oriented independently, yet their separation will constantly have to be monitored and maintained to within a tenth of a micron by laser beams.

The telescopes will be put into an orbit around the Sun a little farther out than the Earth. This is to reduce the background infrared radiation from dust in the inner Solar System that would otherwise drown out the faint signal from the planets. The proposed launch date for TPF is 2012 and for Darwin a little later, though in practice it is quite likely that NASA and ESA will pool their resources into a combined TPF-Darwin mission. Its duration is likely to be six years or more, in which time it should be possible to examine each of the 300 solar-type stars within 50 light-years of the Sun.

The search for extraterrestrial intelligence

In parallel with these programmes, whose goal is to identify planets that support the habitats in which we suspect life can develop, a number of groups are approaching the quest for life from a very different angle. These are the teams that are undertaking a search for extraterrestrial intelligence, or SETI for short. These projects aim to detect signals supposedly being broadcast by intelligent aliens.

The fundamental hypothesis underlying the SETI approach is, of course, the assumption of the existence of intelligent extraterrestrial life. SETI supporters argue that it is now known that many extrasolar planetary systems exist and therefore it is likely that many of these support life. They reason that, at least on some planets, life will develop and evolve in a similar manner to the way it has done so on Earth. They argue that over thousands of millions of years, complex life forms will eventually emerge possessing the faculty of intelligence and that these will ultimately develop an advanced technology and an awareness of their place in the Universe. The next logical step would be for them to try to establish contact with other intelligent beings by signalling their presence.

There are several problems with this line of reasoning, some of which I shall explore further in the next chapter. It rests on a chain of assumptions and unknown probabilities that, in the most favourable circumstances, could indeed lead to the prediction that intelligent extraterrestrial life is widespread in the Universe. On the other hand, the uncertainties are such that a more conservative interpretation can equally well lead to the opposite conclusion, that it is rare or even absent. However, the viewpoint of the SETI enthusiasts is summarised in the following extract from an article that first proposed that a systematic search should be conducted using radio waves, written by Guiseppe Cocconi and Philip Morrison in 1959. 'The probability of success is difficult to estimate, but if we never search, the chance of success is zero.' In other words, if you don't seek, you won't find.

The question then arises of how to set about searching for a message when it is known neither where in the

Universe it might be coming from nor in what form it might appear. Recognising a signal beamed from an extraterrestrial source is not just a question of understanding the language, but a problem of guessing at the very means of communication.

The distances involved are enormous, and the only known methods of sending an interstellar signal are by radio or visible light transmissions. A light beam, perhaps emitted by a powerful laser, is certainly capable of carrying a message. However, assuming it is being sent from an extrasolar planet, it would be hard to pick out against the glare of its central star. For this reason, the radio wavebands are considered more promising; the radio emission from stars is low. This was the basis of Cocconi and Morrison's proposal.

In fact the idea of listening for radio broadcasts from aliens was not strictly speaking a new one. In 1930, in one of the more curious episodes of its kind, the United States Army appealed for a nation-wide radio silence in the hopes that they might detect signals from intelligent life on Mars. It is a measure of how rapidly our knowledge of the Solar System has been transformed that this experiment seems so bizarre today.

Prompted by Cocconi and Morrison, the astronomer Frank Drake carried out the first modern search programme in 1960 using an existing 26-metre diameter radio telescope belonging to the United States National Radio Astronomy Observatory at Green Bank, West Virginia. Pioneers like Drake faced the problem that the radio spectrum is very wide, and there is no way of predicting on what wavelength any broadcasts might be carried. He chose to tune his receiver to 21 centimetres, a wavelength that is emitted universally by interstellar clouds of hydrogen, on the grounds that this would be a sort of natural standard recognised by any technologically advanced life form. He named the programme Project Ozma.

In the small amount of observing time that was available to him, Drake could observe only two target stars; one of them was ε Eridani, the star mentioned above that is now known to possess a giant planet with a highly eccentric orbit. He was unable to detect any meaningful signal from either of them. Although dozens of similar searches were carried out during the 1960s and 1970s at a number of different observatories, they were all limited by time and the equipment available, and none had any more success than Project Ozma.

These failures to discover the slightest signs of extraterrestrial intelligence led to a loss of confidence in both the academic and the political communities, the low point being reached in 1978 when SETI 'won' Senator William Proxmire's Golden Fleece award for wasteful spending of taxpayers' money. The astronomer Carl Sagan campaigned vigorously against this judgement of the project, eventually winning even Proxmire's approval, and as a result a SETI institute founded with congressional money was set up in 1984.

9.6 - The Arecibo radio telescope is built into a natural hollow in the hills of Puerto Rico. Its diameter is 305 metres.

EARTH, LIFE AND THE UNIVERSE

The earlier projects had all suffered from the fact that each could scan only a small number of stars at one or a very few frequencies. There are hundreds of Sun-like stars within a distance of 100 light-years, any one of which is a candidate for SETI. Worse still, the choice of the 21-centimetre wavelength of hydrogen, or wavelengths of other common emitters like hydroxyl at 18 centimetres, depended on the reasoning that rational beings would also be using these wavelengths. It would be far better to be able to scan a wide range of radio wavebands. Indeed it would be worthwhile to listen out on any wavelength between about three and thirty centimetres, a radio region overlapping with microwaves. Outside this waveband, absorption by the atmosphere and background interference from natural sources makes reception difficult. This still leaves a dauntingly large number of frequencies from which to choose.

Modern receivers were becoming capable of receiving several frequencies at once and in 1992 the SETI institute launched an ambitious ten-year monitoring programme named the High Resolution Microwave Survey. Despite the choice of title, which was thought to be politically less sensitive than the Search for Extraterrestrial Intelligence, Congress finally withdrew its support shortly afterwards.

The project survives today on private money, renamed as Project Phoenix and based around one of the largest radio telescopes in the world, the Arecibo dish in Puerto Rico (Fig. 9.6). Phoenix searches the spectrum between 10 and 25 centimetres, broken down into very narrow segments; 2,000 million channels are simultaneously monitored for each target star. There are about a thousand stars on the target list, all within a distance of 200 light-years.

Because Phoenix listens out on so many millions of radio frequencies, an enormous amount of data is continuously being generated. In fact nearly all the 'listening' is done by computers. However, instead of making use of one huge central computer, SETI runs on literally millions of small ones scattered all over the world, personal computers owned by universities, companies and

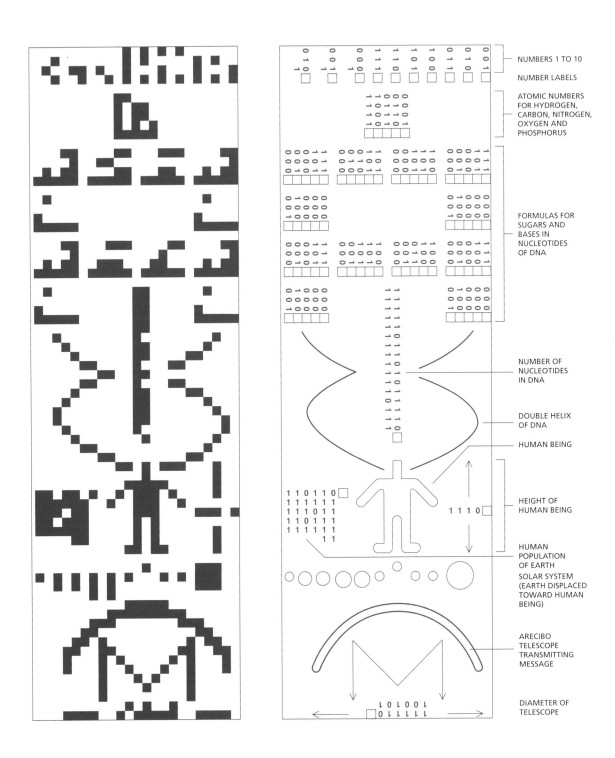

9.7 - The message on the left was broadcast by the Arecibo telescope in 1974. It consists of 1679 elements, arranged into 73 lines and 23 columns (these are both prime numbers, which is intended to help the decoding of the message). It consists, among other things, of depictions of the Arecibo telescope, the Solar System, the DNA helix, a human figure and some of the chemicals of terrestrial life.

private individuals connected to the Internet. The project, called Seti@home, consists of a small program that can be downloaded by anyone and which runs in the gaps when the computer's processor is not otherwise in use. The number of participants, all volunteers, passed the three million mark in 2001 and new members were joining in at the rate of 3,000 a day. It is thought to be the largest computing project in the world.

All these computers are looking for narrow-band signals, that is, signals being emitted close to a single frequency. With easily identifiable exceptions, this kind of broadcast does not occur in nature, and could be the hallmark of a deliberate transmission. If a signal were found to contain obvious non-random patterns, for example repeating sequences or pulses that counted out the whole numbers, it would be confirmation that it was coming from an intelligent source. Nobody expects to find a message encoded into any of our Earth languages, but it might contain arrays of pulses that could be assembled into images or diagrams using the presumably universal terminology of mathematics. A crude example of what might be possible is shown in Figure 9.7. This diagram was actually transmitted into space from the Arecibo telescope to mark its inauguration in 1974. More sophisticated messages have been composed using languages expressed only in mathematical symbols that are built up and defined within the message itself and therefore, it is hoped, universally decipherable.

Communication is a two-way process and it is clearly pointless if everyone in the Universe is waiting and listening for signals if at the same time nobody is sending any. Apart from the occasional deliberate message like the one from Arecibo, we have of course been accidentally transmitting radio signals into space for many years. And, ever since the first BBC television broadcasts in 1936, a steady stream of images has been travelling out into space. Any technically advanced beings within about 60 light-years of us could in principle have received them by now, though they would need antennae and receivers more powerful than Project Phoenix in order to pick them up.

Of course the practice of two-way communication would raise a major problem of delay time. Even if Phoenix were to discover signals from intelligent aliens living at a distance of say 50 light-years from us and were to initiate a dialogue tomorrow, we couldn't expect the earliest reply till around the year 2100.

In practice, despite the efforts of Project Phoenix and other similar undertakings using a variety of radio and optical techniques, not a single signal has yet been received that could be interpreted as having originated from an extraterrestrial intelligence.

10 The nature of extraterrestrial life

The time has come to draw together some of the key points from the earlier chapters and to venture some conclusions about the nature of extraterrestrial life. Because of the present state of our knowledge, the subject matter here is unavoidably more speculative and controversial than the rest of the book. It will certainly be coloured by my personal opinions, although very few of the ideas set out here are completely new. They do, however, reflect my opinion of the most likely directions in which the science of astrobiology will progress. The very title of the chapter is controversial, since nobody yet knows what extraterrestrial life is like. In fact, there is no undisputed evidence that it even exists.

The origin of life – how?

It would be a lot easier to discuss the possible nature of extraterrestrial life if we had a convincing picture of how life originated in the first place. Once it takes a hold, life seems to be tenacious, being able to evolve and adapt to an astoundingly diverse range of environments. All that seems to be needed to keep it thriving, that is, the very minimum requirement, is access to a supply of water and a few commonly occurring chemicals as a source of energy. A supply of solar or stellar energy (and perhaps some protection from harmful radiation) may well encourage the development of more complex forms, but is not essential for the survival of micro-organisms. Even in the absence of energy and water, simple life forms may be able to survive dormant for extended periods, perhaps tens or hundreds of millions of years, and revive when favourable conditions recur.

However, it must have taken more than this for life to materialise from a barren environment. At the very least it would need a supply of organic molecules. Here at any rate we know of an abundant source. Quite large molecules, even amino acids, the building blocks from which proteins are made, occur naturally (they are found in vast quantities in interstellar clouds) and they could have been delivered to the Earth by meteorites or comets. Organic molecules may also have been generated locally by non-biological chemical reactions in hot springs or hydrothermal vents.

Nevertheless, the level of intricacy of an isolated amino acid is quite insignificant in comparison with the complexity of the biochemical structures that are present in even the simplest life forms, such as the proteins, the nucleic acids and the cell membranes. The sequence of events that assembled these from elementary building blocks and the gradual, step by step bridging of this huge complexity gap still fall outside our present understanding. At the moment, we do not know how this transition was achieved, though it is hard to believe that it took place rapidly.

The Earth's fossil record does not extend any further back than 3,500 million years, and we do not know how much earlier than this life appeared on Earth. No fossil-preserving rocks have been found that survive from before this date. The microfossils from this time are bacteria, life forms already possessing and using a complex biochemistry.

It is remarkable that life forms as advanced as this were already present on the Earth no more than 300 million years after the turbulence of the Hadean era. For the most part, the subsequent evolution of life took place excruciatingly slowly. During the next 3,000 million years, until the Cambrian explosion, life remained microscopic. Even to this day, all living things however diverse still use the same biochemical processes as the earliest known micro-organisms. To me, the marvel is not how we developed from bacteria to where we are today, but how the bacteria developed to the point they had reached so soon after the Hadean.

The species that have evolved least from the most recent common ancestor are all thermophilic and this preference for high temperatures must therefore be a characteristic of a phase of early life. However, the earliest fossil bacteria, for example the cyanobacteria, are not thermophiles. A look back at Figure 4.6 shows that they were already the descendants of thermophilic organisms that must have lived at an even earlier time. This could mean that life originated in hot springs or volcanic vents, but it also hints at the possibility that the heat-loving bacterial ancestors existed even *during* the Hadean.

It is believed that the period from about 4,100 to 3,800 million years ago was a period of particularly heavy asteroid impacts. Many large lunar craters are dated to this time and it is assumed that the Earth and the Moon shared a similar history of bombardment. Yet the worst impacts were perhaps tens of millions of years apart. It could be that a few resilient micro-organisms were able to survive each episode, recovering to recolonise the planet. Microbes are individually destructible but entire strains might be so widespread and robust that perhaps not even the largest impacts of this period were able to wipe them out entirely. They could have been forced deep underground, where temperatures are high. Was this the hot environment that fostered the thermophiles?

This early appearance of life on Earth is a considerable mystery. It was already complex 3,500 million years ago and quite probably much earlier, during the late Hadean. There seems to have been far too little time for it to have got to this stage starting with raw materials like amino acids, especially as conditions were presumably even more hostile in the earlier Hadean. Admittedly, since we don't know how the transition to complex cell biochemistry was achieved, we can't say how long was needed for it to take place.

This short time window for the emergence of life on Earth seems to imply one of two things. Either, given the right conditions and raw materials, the conversion of inert raw materials to a state of biological complexity takes place quickly and readily or, alternatively, life arrived already in this advanced state from somewhere else outside the Earth.

In the first case, if it happened so readily on Earth, it must presumably be just as straightforward on any planet passing through a similar phase of its history; and so life will have arisen everywhere in the Universe where the right conditions occur. On the other hand, if life came from outside the Earth, it clearly exists (or at some time did exist) elsewhere in the Universe. Either way, the argument supports the case for the existence or former existence of extraterrestrial life.

The accumulating evidence for the antiquity of life on Earth and the implausibly short timescale available for its development here make it increasingly hard to sustain the first option. I believe the second explanation, that life is of extraterrestrial origin, must be the correct one.

Did life arrive from another planet of the Solar System? Possibly, though it is hardly more likely to have *originated* on any of them than on the Earth. The other planets are not thought to be any older than the Earth and they would also have undergone a simultaneous phase of early bombardment, so the time available for the development of life would not be significantly longer than on Earth. In this respect Mars is perhaps a marginal possibility since its equivalent of the Hadean is likely to have been shorter and less drastic than the Earth's.

Alternatively, perhaps life had an extrasolar origin. At a stroke this would remove the timescale problem since instead of cramming the complete emergence of life into a brief 100 million years or so, there would be the luxury of the whole of the previous lifetime of the Galaxy of which to take advantage; 10,000 million years or more. This does not in any way help to explain how life originated but at least it provides a much more credible length of time in which it could have managed so to do. How plausible is this?

The origin of life – where?

The fact that meteorites from Mars and other bodies have reached the surface of the Earth suggests a mechanism by which life could be imported from or exported to other planets of the Solar System. The Martian meteorites were originally surface rocks thought to have been ejectled by an asteroid impact. If there were microbes lodged in the rocks, they could have survived the interplanetary journey to Earth. Terrestrial bacteria have survived unprotected for periods of years on spacecraft and could possibly remain dormant for millions of years inside a meteorite. However, could the microbes have survived the shock of the blast-off from Mars or the intense heating of the passage through the Earth's atmosphere at 40,000 kilometres per hour?

Examination of the Martian meteorite ALH84001 shows that during its plunge to Earth it reached temperatures of 12,000°C at its surface, enough to melt its outer layers. However, rock is a poor conductor and the descent would have been extremely rapid, and its interior never rose above a very comfortable 40°C.

Nor is the trauma of the expulsion from the Martian surface likely to have been a problem. A team led by Mark Burchell of the University of Kent at Canterbury in England experimented with samples of *Rhodococcus rhodochrous*, which normally lives in oceanic vents at a depth of five kilometres below the surface. To simulate the shock of the ejection, these bacteria were fired at speeds of 18,000 kilometres an hour into targets of granite and chalk. They survived.

The microstructures found within ALH84001 may or may not turn out to be biological in origin, but if there *had* been any microbial life embedded inside, there seems to be no reason why it could not have withstood the rigours of interplanetary space travel. The exchange of life like this between Earth and Mars, and perhaps other planets too, is perfectly feasible.

The question of whether life might have come from farther afield, from outside the Solar System, is much more controversial. Hitching a ride in rocks like Martian meteorites to travel between the stars is out of the question, since rocks blasted from the surface of a planet could never reach a speed high enough to escape from their own stellar system. Yet we know that material does travel from star to star. At the end of their lives stars expel large amounts of matter into space that mixes in with the existing interstellar gas and dust clouds, and this is the material from which new stars and planets form. These clouds contain quite complex organic molecules. Could they also contain micro-organisms, perhaps the last survivors from planets that used to orbit dying stars?

It is highly likely that the Earth has been sprinkled with organic molecules carried by comets, formed out of the primeval material of the Solar System. Perhaps this material also contains primitive life that originated in extrasolar systems. Perhaps life itself is part of the cosmic cycle.

Is there really any chance that micro-organisms could remain viable during a lengthy interstellar journey as part of a gas cloud? Unshielded by a rocky covering, wouldn't they be destroyed by the ultraviolet and other high-energy radiation that permeates space?

The tolerance of microbial organisms to radiation is very variable. Some bacteria are incredibly robust; for example, *Deinococcus radiodurans* can withstand radiation thousands of times greater than other organisms and the bacterium *Micrococcus radiophilus* has an especially effective mechanism for repairing X-ray damage to its DNA. In an experiment designed to replicate the conditions of interstellar space, Peter Weber and Mayo Greenberg of the University of Leiden in the Netherlands irradiated bacterial spores with the equivalent of 2,500 years exposure to ultraviolet starlight. Although 99.9% were killed, a small number did survive. In any case, interstellar travellers would not need to depend solely on their own protective mechanisms. They would become coated in the dusty, sooty material that accompanies the gas thrown off by old stars. Coatings only a micron thick would offer almost total shielding from ultraviolet radiation.

The journey time between the stars may not be unrealistically long. Material in planetary nebulae expelled from dying stars can be accelerated to speeds of over 1,000 kilometres per second and can mingle with the existing interstellar clouds within 100,000 years. It would be a matter of chance how long it took before these clouds formed into new stellar systems, but it is not impossible for the cycle to be completed within a few million years. Moreover, terrestrial bacteria are known that have survived in a dormant state for tens and possibly hundreds of millions of years

In this way, dormant organisms could have spread from star to star through the Galaxy as a natural part of the cosmic cycle. The idea that life may propagate widely throughout space in the form of spores is not new. Svante Arrhenius (the same Swedish chemist who suggested the connection between carbon dioxide and global warming) proposed this theory and gave it the name panspermia as far back as 1908. In its modern form it has been elaborated and championed by Fred Hoyle and Chandra Wickramasinge of Cardiff University in Wales. In its support they point out that many of the properties associated with certain bacteria, such as tolerance to extreme environmental conditions, are not of a kind that one would have expected to evolve on Earth.

If this theory is correct, life must be spread wide across the Galaxy in the form of dormant organisms carried between the stars by interstellar clouds. Wherever stars and planetary systems form, comets form too, from the same cloud material. The comets preserve the life forms and scatter them on to the planets where they either perish or flourish, depending on the conditions prevailing. Life is highly adaptable, needing only simple chemicals, energy sources and water to develop, and if extrasolar planets that can provide a suitable environment are common, life on extrasolar planets must be common too. Is it possible to imagine what it might be like?

The nature of extraterrestrial life

Life on extrasolar planets is most likely to be present as relatively simple organisms like bacteria. These are of course the earliest known life forms to have occurred on the Earth and they are to this day still the commonest, the most pervasive and the most successful organisms ever to have inhabited the planet.

It is quite likely that life on most planets never develops beyond this stage. Some planets will be too small or will be insufficiently geologically active to retain their atmosphere and oceans, like Mars. Some will lose all their liquid water because of a runaway greenhouse effect, as on Venus. If life emerges on these, it may either die out relatively quickly or else remain in a simple or dormant phase.

Other planets will be more like the Earth, preserving their oceans and atmospheres for long periods of time. Even on these it may not be inevitable that life proceeds to become more diverse and complex. It is possible to imagine circumstances that might prevent this, such as a much higher rate of meteoric bombardment than on the Earth. On worlds like this life may persist, but remain simple.

Some astronomers have argued that the Earth is exceptional and that complex life is exceedingly rare in the Universe. They point out that the Earth is in a stable orbit at just the right distance from the Sun, that it has enough mass to retain its atmosphere and oceans and that it continues to generate enough internal heat to keep its geological and Earth system cycles going. They also contend that the incidence of catastrophic impacts is low and attribute this to the lucky existence of our giant neighbour Jupiter, which acts as a shield, attracting and sweeping up incoming asteroids and comets that would otherwise collide with the Earth.

I find this line of reasoning unconvincing. These circumstances do not seem so exceptional and there appears to be no reason why they could not prevail in huge numbers of extrasolar planetary systems. Giant planets, for example, are common. However, it is certainly true that there is nothing inevitable about the emergence of complex life from simple forms, and the latter will be far more widespread. Nor do I suppose that the circumstances in which the Earth has evolved will be exactly repeated elsewhere. As a result, where complex life has emerged on other planets, it will have developed in distinct and unpredictable ways, the product of the particular hazards it has met along the way.

We can imagine that on a planet where it survives for a sufficiently long time, life diversifies and develops a complex interdependent relationship with its planet, as it has in our own Earth system. However, the course of evolution will not take the same path there as on Earth. The life forms that emerge will be the result of innumerable chance events such as the changes in the environment and the incidence of random catastrophic accidents. Each mass extinction event, as on Earth, resets the clock and evolution takes off anew in a different direction. Just as the existence of humans on Earth is the result of chance rather than a predestined progression, the emergence of trees, mammals or insects on another planet cannot be expected.

This raises the question: is the emergence of intelligence inevitable? Is it a property of any increasingly complex and diverse biosphere that, given enough time, will always arise out of this complexity? Is it the case that any planet supporting such a biosphere will develop intelligent life? If a natural (or self-induced) catastrophe were to wipe out all animals including humans, simpler life forms would repopulate the Earth, but would intelligence then develop in another, as yet unborn species?

Does intelligence matter? It is a distinguishing feature of our species and we tend to accord it a high importance. This is certainly justifiable in a cultural connotation, but a very self-centred view in biological terms. If we believe that intelligence is important, then our species is automatically the most important one! It is not even clear what its usefulness is in evolutionary terms, that is, whether it confers a lasting evolutionary advantage. If it does, why doesn't it occur more often? After all, the eye evolved independently in many different species on Earth, but intelligence of a human kind appears to have emerged only once.

Our own species had already successfully emerged long before we became intelligent, or at least before we started to make extensive use of our brains. This occurred only about 40,000 years ago, when there was a remarkable growth in speech, art and technology. It may well be that intelligence has given us an edge over some other species in the subsequent competition for survival on the planet. Yet it is evidently not necessary for endurance; in this respect the bacteria would be judged far more successful than we are. There is also the danger that this very intelligence, or the technology it has spawned, could lead to long-term damage to the planet and a threat to our continued existence on it.

If this picture is correct, life is widespread and perhaps universal but planets supporting complex life systems like the Earth system are probably rather uncommon. Even on these planets, the life forms are unlike those on Earth and intelligent life is rare.

This conclusion may seem to run counter to the reasoning that we might call the argument of overwhelming statistics, outlined in Chapter 9. This says that there are a countless number of stellar systems in the Galaxy, so even if only a fraction of them accommodate terrestrial planets (or other potential biospheres such as Europa) and only a fraction of these harbour life, and if only a fraction of the ones that support life develop intelligent life forms, this still means that intelligent life is common.

Astronomy has shown us that the reasoning is plausible (though not yet proven) as far as the existence of large numbers of potentially life-supporting planets is concerned. However, the statistical argument cannot be carried through into the domain of biology. The chance of some form of life evolving is very different from the chance of the evolution of a particular form of life. We cannot predict the path that evolution will take and cannot expect that it will be anything like on Earth. Specifically, we cannot argue that intelligence is widespread.

If intelligent life is sparse, the number of extraterrestrial civilisations close to us in space will be low, raising questions about the likely success of SETI. Moreover, there is not only a proximity-in-space problem, there is a proximity-in-time problem too. Even if intelligent life were to emerge in a nearby stellar system, say ϵ Eridani, there is only the remotest possibility of it being at the same level of technical development as ourselves at the same time. The entire history of our species has lasted only a minute fraction of the age of the Galaxy and we have had the technology to receive radio signals from space for a mere 50 years or so. Even if the ϵ Eridanians transmitted continuously in our direction for a million years, nobody on Earth would have been able to receive their broadcast if the peak of their civilisation had coincided with the Jurassic period on Earth.

However, the problem is even more difficult than this. There is also the problem of coincidence at an intellectual level. Underlying SETI is the assumption that an alien life form has developed not only the same technology but also the same aspirations and motivations as ourselves, in particular the desire to initiate interstellar communication. It seems to me that this concept of intelligence is too human-centred. We all believe we are more intelligent than other animals and we all tend to say an animal is more intelligent if it is more human-like. Yet intelligence is not a single quantifiable attribute, but a whole collection of processes and skills that can be used to solve a range of problems. Animals are cleverer at certain tasks and less clever at others.

We can communicate only in the most limited way with a very few of the species that share our own planet, to whom we are most certainly more intimately related than to any inhabitant of another one. Indeed, it is at times hard enough to establish clear and unambiguous communication between different cultures within our own species. To me, it seems a completely unjustified assumption to think we might share the same kind of intelligence with an extraterrestrial life form. For these reasons I am pessimistic about the chances of SETI achieving a positive result in the foreseeable future.

Conclusion

In a laboratory at NASA's Ames Research Center in California, scientists have carefully recreated a tiny volume of interstellar space. Inside a vacuum chamber, minute amounts of common gases like water vapour, methane, ammonia and carbon monoxide can be cooled to temperatures of −260°C and bathed in ultraviolet light. These are the kinds of conditions that are prevalent in the interstellar clouds. The intensity of radiation is such that each hour in the chamber is equivalent to a thousand years in space.

Solid surfaces can be placed in the chamber to simulate the dusty grains that are also present in the clouds. Molecules of gas freeze on to these surfaces and build up into thin layers of ice. New chemical compounds form and a profusion of organic molecules appears, even small quantities of amino acids, just like the ones detected in actual interstellar clouds.

When Lou Allamandola from the NASA Ames team and David Deamer of the University of California at Santa Cruz carried out this experiment, they were pleased with the confirmation that quite complex chemical reactions can take place in space. However, what they observed next was a complete surprise. When they dipped the new compounds in water, they assembled themselves into droplets, tiny capsules about 10 to 40 microns across, with the organic molecules sheltered in their interior. While nobody has suggested that these structures are alive, they do have a resemblance to living cells. If compounds like this were gathered up by comets and delivered to ocean-bearing planets, they could have been a key step in the creation of the very first forms of life.

For the first time in our history, using experiments like these, the subject of extraterrestrial life is being moved into the area of testable theory. Speculation is all very well, but theories must be rigorously proven before they become science. So how might we test some of the ideas in this chapter? What sort of results might we hope to see in the coming years or in the longer term? Here are some predictions.

We are reasonably sure that the exchange of life between Solar System planets is possible and we believe that Mars was habitable in the distant past. I predict that fossil life will be confirmed on Mars, though the discovery will probably have to wait until Martian rocks have been retrieved and returned to Earth, perhaps around the year 2015. If we were ever to find fossil Martian organisms, we would want to establish how closely related they were to terrestrial life forms and whether we could place our ancestries on a common tree of life. Martian fossils might even hold important clues about the nature of pre-bacterial life. Life on Mars is almost certainly extinct though there is an outside chance that it survives in a dormant state. In this case it might be possible to discover the genetic and metabolic processes it used; are they, for example, based on DNA and proteins?

At about this time, by the middle of the second decade of the twenty-first century, many more extrasolar planets should have been discovered, both giant and Earth-sized. At that point I expect that a NASA/ESA giant interferometric telescope will detect the first extrasolar planets with atmospheres bearing the imprint of an active biosphere, and this will be the first proof that life exists outside the Solar System. Regrettably, it seems that for the foreseeable future we will be unable to obtain much evidence about the form that this life takes or its relationship to us, evidence that could provide support for the panspermia theory.

My next prediction is a negative one, and it could be disproved at any time without warning. I do not think that the SETI project will ever succeed in detecting any signals from an obviously intelligent alien source. Despite my conviction, I do not think that this is a reason to give up the search. I agree that if you don't seek, you won't find. For my part, I think it is highly unlikely that the project will be successful in its aims and I am happy that the undertaking is a private one and not a call on public funds.

Finally, I am certain that living extraterrestrial organisms will eventually be discovered somewhere. This is of course the easiest prediction of all to make, since it can never be disproved as long as the search goes on. However, if the discovery does not take place in our Solar System then it will be a very long time indeed before it comes to pass.

I will conclude with this thought. I have described how comets are likely to be the carriers of complex molecules and possibly also cell-like capsules full of organic compounds originating in the interstellar clouds. I have suggested that they may even carry spore-like cells that have been picked up from clouds that have crossed the spaces between the stars. When a comet sweeps into the inner Solar System, its interior is shielded from the worst of the solar radiation, but perhaps it warms enough to melt a little ice, producing water, the essential to life. Perhaps, just perhaps, the nearest viable life forms to us are to be found in the comets.

Two exciting cometary space missions are planned in the near future. NASA's Stardust spacecraft is due to rendezvous with comet Wild 2 in January 2004, when it will collect samples of the material surrounding the comet and return them to Earth in 2006. It will be followed by ESA's ambitious Rosetta mission, whose main objective is to intercept and go into orbit around comet Wirtanen. Rosetta will carry a landing craft and set it down onto the solid surface of the comet's nucleus in August 2012. Will one of these missions be the first to encounter extraterrestrial life? My final prediction, that living extraterrestrial organisms will eventually be discovered somewhere, may just turn out to be the first to be fulfilled.

Geological periods and eras since the Cambrian

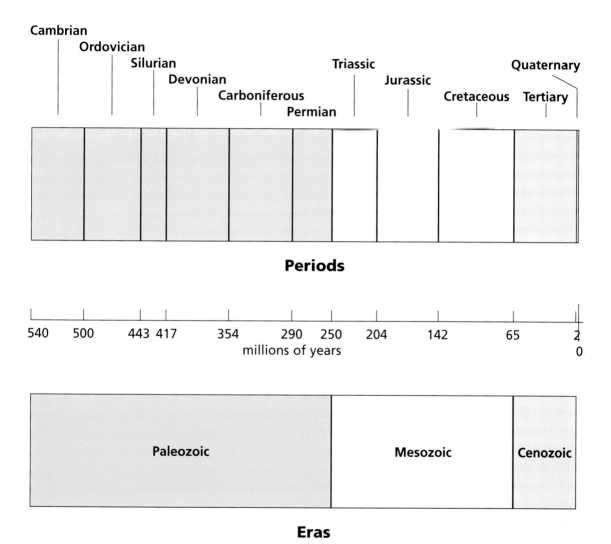

Diagram showing the division of geological time since the start of the Cambrian into periods and eras. Compare this with Figure 3.4.

Glossary

$$H_2N - \overset{\displaystyle H}{\underset{\displaystyle H}{\overset{|}{\underset{|}{C}}}} - COOH$$

glycine

$$H_2N - \overset{\displaystyle H}{\underset{\displaystyle CH_3}{\overset{|}{\underset{|}{C}}}} - COOH$$

alanine

$$H_2N - \overset{\displaystyle H}{\underset{\displaystyle CH_2}{\overset{|}{\underset{|}{C}}}} - COOH$$
$$OH$$

serine

$$H_2N - \overset{\displaystyle H}{\underset{\displaystyle CH_2}{\overset{|}{\underset{|}{C}}}} - COOH$$
$$SH$$

cysteine

G.1 - *The chemical structure of some typical amino acids. All amino acids consist of a hydrogen, an amino group (–NH₂), a carboxyl group (–COOH) and a side chain which is unique to each amino acid.*

Amino acid. The amino acids are a group of relatively simple organic molecules, compounds of carbon, hydrogen, nitrogen and oxygen, sometimes also containing sulphur or iodine (see Fig. G.1). Proteins are formed from linear chains of amino acids. More than 100 amino acids occur in nature, particularly in plants, but the vast majority of the proteins found in living organisms are composed of only 20 different kinds.

Archaea. Along with the bacteria and eucarya, one of the three domains into which all life forms are classified (see Fig. 4.5). Archaea are very ancient prokaryotic life forms usually found in naturally hot environments, such as hot springs, volcanic vents and boiling mud pools.

Asteroid. The asteroids, also known as the minor planets, are a swarm of rocky or metallic bodies whose orbits mostly lie between Mars and Jupiter. The largest is a little over 900 kilometres in diameter, but most are very much smaller, irregularly shaped and only a few kilometres across.

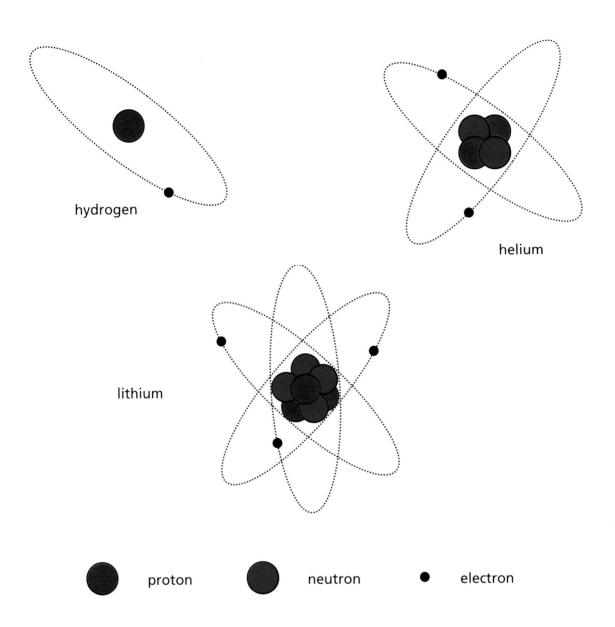

G.2 - *The structure of the three lightest elements, hydrogen, helium and lithium. The nucleus, composed of protons and neutrons, is surrounded by one or more electrons. Only the commonest isotopes are shown.*

Atom. The atom is a basic unit of matter, which can be imagined as a nucleus surrounded by a cloud of electrons. Electrons are very light particles and so nearly all the mass in an atom is concentrated into its nucleus, which is a tightly packed assembly of particles called protons and neutrons. The number of protons in the nucleus of an atom is always given by the atomic number. The number of neutrons in the nucleus can vary, but is about the same as and usually rather more than the number of protons. Figure G.2 shows the atomic structure of the three lightest elements, hydrogen, helium and lithium.

ATP (adenosine triphosphate). All organisms make use of the energy-transferring compound ATP. It is the common energy currency of every cell. ATP is formed when the cell acquires energy and it is broken down again at the point where the cell needs to use it up again. It is the mechanism by which biologically useful energy is transferred from energy-producing to energy-consuming reactions.

Autotroph. Organisms that acquire carbon from the inorganic compound carbon dioxide (CO_2) are called autotrophs. The green plants and some bacteria are autotrophic.

Bacteria. Along with the archaea and eucarya, one of the three domains into which all life forms are classified (see Fig. 4.5). Bacteria are very ancient prokaryotic life forms. They are very resilient and inhabit a wide range of environments, including soil, rock, water and the bodies of animals.

Base, base-pair. See nucleotide base.

Brown dwarf. Brown dwarfs are stars that are so small (less than about 8% of the mass of the Sun) that when they formed they never achieved temperatures high enough for normal nuclear fusion reactions to start up.

Carbonaceous chondrite. A type of meteorite having a composition that matches the Sun's very closely with the exception of the more volatile materials. They were formed from the primordial material that condensed out of the solar nebula.

Carbonate. Carbonates are chemical compounds derived from carbonic acid (H_2CO_3), which is formed when carbon dioxide dissolves in water.

Chemoautotroph. Organisms that extract energy from purely inorganic compounds are a type of autotroph known as the chemoautotrophs. Chemoautotrophs can exist in the absence of both light and organic food.

Chirality. A chiral molecule is a molecule that exists in two forms that are mirror images of each other in three dimensions. The amino acids that occur in proteins are chiral, belonging to the L-form. Their mirror images, the D-amino acids, are not used biologically.

Chromosome. In eukaryotes, the DNA is tightly wound and folded into tiny packages called chromosomes. Almost every human cell preserves an identical set of 23 pairs of chromosomes.

Codon. The four nucleotide bases A, C, G and T are used to carry the instructions for making all organisms. The order of these letters provides the code. Within a gene, each adjacent triplet of bases, called a codon, corresponds to a single amino acid.

Comet. Comets are amongst the smaller members of the Solar System. They are irregular clumps of ice and dust typically a few kilometres across that travel round the Sun in highly elliptical orbits. They are icy remnants of the solar nebula that condensed a long way from the Sun.

Cytoplasm. In a cell, the cytoplasm is the watery liquid enclosed by the cell membrane and (in eukaryotes) exterior to the nucleus. It contains a mixture of large organic molecules such as proteins and nucleic acids.

Deuterium and tritium. Deuterium and tritium are two isotopes of hydrogen sometimes referred to as heavy hydrogen. A deuterium nucleus has one proton and one neutron; tritium has one proton and two neutrons.

DNA (deoxyribonucleic acid). A DNA molecule is composed of two long chains of nucleotides that coil around each other in a double helix. Each chain forms a backbone, and sticking out from the side of each nucleotide is a subunit called a base. The DNA nucleotide bases are cytosine, thymine, adenine and guanine, abbreviated to C, T, A and G. The overall structure is like a twisted ladder – the sides are made up of sugars and phosphates and each rung is formed out of one base-pair. Each A on one chain pairs with a T on the other, forming an A-T ladder rung, and each C on one chain pairs with a G on the other. DNA stores the genetic information that determines the essential character and hereditary properties of an organism, using the genetic code. DNA is also the means of transmitting the information from generation to generation.

Doppler shift. The change in the frequency or wavelength with which radiation reaches an observer as a result of the motion of the source towards or away from the observer. If the source is approaching the observer, the frequency rises and the wavelength becomes shorter (a blue shift); if the source is receding from the observer, the frequency drops and the wavelength becomes longer (a red shift).

Electron. An elementary particle that carries a unit of negative electrical charge. The chemical properties of an atom depend on the arrangement of the electrons making up the cloud that surrounds the nucleus. When atoms combine into molecules they form a shared electron cloud.

ESA. The European Space Agency.

Eucarya, eukaryote. Eukaryotes can be either single-celled, like the amoeba, or multi-celled, but all are much larger than the prokaryotes. Each eukaryotic cell contains a nucleus that holds the DNA. After the bacteria and the archaea the eukaryotes form the third domain of life, called the eucarya. All multi-celled organisms belong to this domain, including the fungi, plants and animals.

Gene. A gene is a block of DNA instructions that specifies the complete assembly sequence of one protein. The Human Genome Project identified 30,000 – 40,000 genes in humans, though the true number could be much higher.

Genetic code. Three adjacent DNA nucleotide bases form a codon, which codes for an amino acid. For example the sequence C-A-U represents an amino acid called histidine and the sequence G-U-C represents one called valine. Since there are four bases, there are 64 possible codons, three of which do not code for amino acids but indicate the end of a sequence.

Genome. The genome of an organism is the complete sequence of bases carried on its DNA. The human genome sequence is 3,100,000,000 base-pairs long.

Giant (or gas giant) planet. The giant planets Jupiter, Saturn, Uranus and Neptune inhabit the outer regions of the Solar System. Their solid cores are buried deep beneath vast gaseous atmospheres.

Glacial and interglacial. The periods of advance and retreat of the ice sheets covering northern Europe and North America are called glacials and interglacials. They are presently occurring on a timescale of about 100,000 years and the last retreat took place about 10,000 years ago. This cycle of glacials and interglacials, which has continued for about two million years, is taking place within the Ice Age in which the Earth still finds itself today.

Greenhouse effect, greenhouse gas. A greenhouse gas is a gas in a planetary atmosphere that contributes to the greenhouse effect. Although these gases are transparent to the incoming light of the Sun, they block and absorb much of the outgoing infrared radiation from the surface of the Earth, with the result that the atmosphere, and hence the surface of the planet, warms up. Carbon dioxide, water vapour and methane are important greenhouse gases.

Habitable zone. A star's habitable zone lies in the range of distances from the star within which a planet has a surface temperature between the freezing and boiling points of water. Within the Solar System, Venus lies near the inner, hotter boundary of the zone and Mars lies near the outer, colder boundary. The Earth lies comfortably within the zone. Hotter stars have more distant, extended habitable zones and cooler stars will have closer, narrower zones.

Heterotroph. Organisms that acquire carbon by feeding on organic matter, such as sugars, proteins, fats, or amino acids, are called heterotrophs. Humans, other animals and many bacteria are heterotrophic.

Ice Age. Ice Ages are periods lasting several million years when extensive areas of ice cover the Earth. The Earth is currently in an Ice Age now. Other major Ice Ages have occurred a number of times in the past, for example towards the end of the Precambrian, between 700 and 800 million years ago, and again about 280 million years ago near the boundary of the Carboniferous and Permian periods. However, the Ice Ages are relatively rare periods of geological history. During the 90% of the time that the Earth has not been in an Ice Age, mean temperatures were up to about 15°C degrees higher than at present. (The term ice age, with lower case initial letters, is sometimes also used to refer to the much shorter glacial periods within the present Quaternary Ice Age.)

Interglacial. See glacial and interglacial.

Isotope. Atoms whose nuclei contain the same number of protons but a different number of neutrons are called isotopes of the same chemical element. For example, the nucleus of the commonest isotope of carbon is carbon-12, which contains six protons and six neutrons. About 99% of naturally occurring carbon is of this form but about 1% is in the form of carbon-13, with six protons but seven neutrons. Another example is uranium. Uranium-238 is the commonest isotope, with 92 protons and 146 neutrons, but again about 1% is in the form of uranium-235, with 92 protons but only 143 neutrons. Isotopes of the same element have slightly different masses but essentially the same chemical properties.

Light-year. A unit of distance used in astronomy. It is the distance travelled by light in one year. The speed of light is 300,000 kilometres per second and a light-year is about 9,460,000,000,000 kilometres. The nearest star is about 4.2 light-years away.

Major planet. There are nine major planets in the Solar System. In order from the Sun, they are Mercury, Venus, Earth, Mars, Jupiter, Saturn, Uranus, Neptune and Pluto.

Meteorite. A rocky or metallic body of extraterrestrial origin that has penetrated the Earth's atmosphere and fallen to its surface. Although the very largest are capable of excavating impact craters, these are extremely rare. Most are only a few centimetres in size.

Micron. One micron is a length of 1/1000th of a millimetre. The word is a contraction of micromillimetre, i.e. a millionth of a metre.

Minor planet. See asteroid.

Molecule. A molecule consists of a combination of atoms that are chemically bound to each other. Some simple compounds mentioned in the text include oxygen (O_2), carbon dioxide (CO_2), water (H_2O) and methane (CH_4). When atoms combine in this way, the properties of the compound are quite different from those of the constituent elements.

Most recent common ancestor. Various lines of argument suggest that all present day life forms are cousins to each other and that all life on Earth has a common origin. On the tree of life, the most recent common ancestor is located immediately before the very first branching. However, it would not be the first common ancestor, since this organism presumably evolved from something else.

Mutation. An alteration to the genetic material of a cell that is propagated during cell division. Mutation may arise as a DNA copying error or as a result of external factors such as radiation or exposure to toxic chemicals.

NASA. The National Aeronautics and Space Administration, the space agency of the United States.

Neutron and proton. Neutrons and protons are elementary particles that bond in various proportions to form the various atomic nuclei. They are of approximately equal mass (about 1800 times greater than that of the electron) and make up nearly all of the mass of an atom. Protons carry a small positive electrical charge (equal and opposite to that of the electron) whereas neutrons are electrically neutral.

Nucleic acid. There are two types of nucleic acids, ribonucleic acids (RNA) and deoxyribonucleic acids (DNA). Both are long chain molecules consisting of repeating units called nucleotides (Fig. G.3).

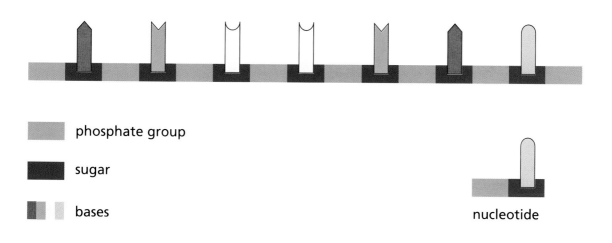

G.3 - *Nucleic acids are long chains of nucleotides. Each nucleotide consists of a sugar and a phosphate group forming the backbone and a nucleotide base to the side.*

Nucleotide. An organic compound comprising a nitrogen-containing base linked to a sugar and a phosphate group. The nucleotides are the building blocks of nucleic acids.

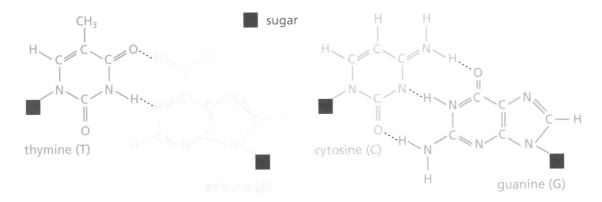

G.4 - *The chemical structure of the nucleotide bases thymine, adenine, cytosine and guanine.*

Nucleotide base. A component of a nucleotide, a compound of carbon, hydrogen, nitrogen and oxygen. In DNA, the nucleotide bases are cytosine, thymine, adenine and guanine (C, T, A and G). An A on one chain bonds to a T on the other forming an A-T base pair and a C bonded to a G forms a C-G base pair. In RNA the bases are C, U, A and G – the same as DNA but with uracil (U) substituting for thymine.

Organic molecule. Molecules based on carbon are called organic molecules (though very simple molecules like carbon dioxide (CO_2) and the element carbon itself (C) are not considered to be organic). They are not necessarily produced by biological processes: the use of the word organic is accounted for by the fact that carbon plays such a key role in the chemistry of life.

Panspermia. The name given to the theory that life propagates itself widely throughout interstellar space in the form of dormant spores. It was first proposed in this form by the Swedish chemist Svante Arrhenius in 1908.

Photosynthesis. The process by which an organism converts light energy into chemical energy. Sunlight is used to convert carbon dioxide and water to carbohydrates and oxygen; free oxygen is released as a by-product.

Phototroph. Organisms that use photosynthesis to generate cellular energy from light energy are a type of autotroph known as the phototrophs. Plants are the most obvious examples of phototrophs, but many bacteria, like the cyanobacteria, are also phototrophic.

Planetary nebula. Towards the ends of their lives, less massive stars like the Sun expel great quantities of gas from their surfaces in shells called planetary nebulae.

Planetesimal. A small body formed out of the material of the solar nebula. The major planets are thought to have been built up from planetesimals by a process of accretion.

Plasma. In astronomy, a plasma is a special state of matter, often but not always at high temperatures, such as in the interiors of stars. In most ways a plasma behaves like an ordinary gas but it is a good conductor of electricity and it is also susceptible to magnetic fields.

Prion. A prion is an aberrant form of an otherwise harmless brain cell protein that can be acquired by infection or inherited from a mutation. It spreads by warping healthy proteins, inducing them to change shape and distort into the abnormal form. It is responsible for a variety of degenerative brain diseases such as bovine spongiform encephalopathy (BSE) in cattle and Creutzfeldt-Jakob disease (CJD) in humans.

Prokaryote. A microscopic, single-celled organism containing no nucleus. There are two domains of prokaryote, the archaea and the true bacteria.

Protein. Proteins perform a wide range of vital functions in living things. For example enzymes control the metabolism of the organism and hormones act as chemical messengers between different parts of the body. Other proteins transport material from one part of the body to another, protect the body from disease and form structural components of the organism like collagen. Proteins are formed by a linear arrangement of amino acids assembled in a particular order.

Proton. See neutron.

Respiration. A process used by organisms to extract energy from organic compounds. The compounds are broken down into carbon dioxide and water and free oxygen is used up in the process.

Ribosome. Ribosomes are the sites of protein synthesis in cells, composed of RNA and proteins.

RNA (ribonucleic acid). RNA consists of a single strand of typically hundreds or thousands of nucleotides. The nucleotide bases in RNA are adenine, guanine, cytosine, and uracil. There are three main types of RNA: messenger RNA (mRNA), transfer RNA (tRNA) and ribosomal RNA (rRNA). In protein formation, mRNA carries a copy of the DNA in the nucleus to the ribosomes, tRNA transports the amino acids to the ribosomes and the amino acids are linked into proteins by rRNA.

Satellite. Whereas a planet is a body that orbits the Sun, a satellite is a body (natural or artificial) that orbits a planet. The Moon is a satellite of the Earth.

Sedimentary rock. Sedimentary rocks are formed from eroded older rock particles, soil, and dissolved minerals that were transported by wind or water and then deposited as layers of sediment which subsequently solidified. Fossils can often be found in sedimentary rocks.

SETI. The search for extraterrestrial intelligence.

Solar nebula. The gaseous cloud from which the Sun and planets formed. As the cloud contracted under its own gravity, the central region grew very dense and hot, eventually evolving into the Sun. The rest of the cloud flattened into a rotating disc. The planets and the rest of the Solar System condensed out of the material of the disc.

Supernova. The final stage of the life of a massive star is an enormous explosion called a supernova. Most of the material of the star is ejected out into space as a result.

Terrestrial planet. The terrestrial planets Mercury, Venus, Earth and Mars are rocky bodies with iron-rich cores and silicate mantles. Some of the larger satellites, such as Io and the Moon, have a similar structure.

Thermophile. Thermophiles are prokaryotes that prefer an environment in which the temperature is at or above 70°C. Most archaea and some bacteria are thermophilic.

Tonne. One tonne is a mass of 1,000 kilograms, sometimes known as a metric ton.

TPF. Terrestrial Planet Finder, a space telescope planned by NASA as part of its *Origins* programme. Its mission will be to search for life-sustaining planets.

Tritium. See deuterium.

Volatile. A volatile substance is one that solidifies at a relatively low temperature, such as ice, or remains a gas, like hydrogen or helium. Substances that solidify at higher temperatures, like metals and minerals, are called refractory. Bodies that formed in the inner Solar System are relatively rich in refractory materials and poor in volatiles.

Further reading

This is a deliberately short selection of books that go more deeply into many of the topics covered by *Earth, Life and the Universe*. They have been chosen for their readability and for their recent (at the time of writing) publication dates. Many of them have their own more extensive bibliographies.

Part I (The Earth and the Universe)

Unfolding Our Universe by Iain Nicolson (Cambridge University Press 1999). ISBN: 0521592704.

The New Solar System by J Kelly Beatty, Carolyn Collins Petersen and Andrew Chaikin (Sky Publishing Corporation and Cambridge University Press, 4th edition 1999). ISBN: 0521645875.

Part II (Earth and Life)

The Spark of Life by Christopher Wills and Jeffrey Bada (Oxford University Press 2001). ISBN: 0198662769.

Stepping Stones by Stephen Drury (Oxford University Press 2001). ISBN: 0198508077.

Earth Story by Simon Lamb and David Sington (BBC Books 1998). ISBN: 0563387998.

Earth: Evolution of a Habitable World by Jonathan I Lunine (Cambridge University Press 1998). ISBN: 0521644232.

The Fifth Miracle by Paul Davies (Penguin Books 1999). ISBN: 0140282262.

Part III (Life and the Universe)

The Search for Life on Other Planets by Bruce Jakosky (Cambridge University Press 1998). ISBN: 0521598370.

Life on Other Worlds and How to Find It by Stuart Clark (Springer-Verlag 2000). ISBN: 185233097X.

Publisher's acknowledgements

The Publishers acknowledge the generosity of the individuals, corporations and institutions which have permitted their images and other material to provide the many illustrations in this book and which have afforded such dramatic effect to the author's text. The appropriate credits are as follows:

Fig. 1.1	Courtesy of Michael Reichmann from his *Luminous-Landscape* website
Figs. 1.2 2.1 2.3 3.4 4.5 6.2 6.3 6.4 6.5 9.2 9.3 9.4 G.0 G.1 G.2 G.3 G.4	Illustrations all drawn by Tony Stanger of Premier Printers Limited from sketches provided by the author
Fig. 1.3	Courtesy of National Hydrogen Association
Fig. 1.4	Courtesy of SOHO/ESA/NASA
Fig. 1.5 1.6	© Anglo - Australian Observatory and photographs by David Malin
Fig. 1.7 (and cover)	Courtesy of European Southern Observatory
Fig. 2.2	© 1995 California Institute of Technology. US Government sponsorship under NASA contract NAS27 - 1270
Figs. 2.4 2.5 3.1 3.2 7.7 8.4 8.5 8.6 8.7 8.8	Courtesy of NASA Johnson Space Center, Houston, Texas
Fig. 2.6	© John Laborde
Figs. 2.7 2.8 9.1	Courtesy of Hubble Space Telescope - material created with support to AURA/STScl from NASA contract NAS5 - 26555
Fig. 3.2 Inset	Courtesy of Discover West Holidays © Kalhaven Holdings Pty Ltd

Fig. 3.3	Courtesy of IGPP Center for Astrobiology
Figs. 4.1 4.2 4.3 4.4	Redrawn from illustrations contributed by Darryl Leja and the National Human Genome Research Institute to Access Excellence sponsored by Genentech of San Francisco
Fig. 4.6	Redrawn from an illustration from *Evolution of Hydrothermal Ecosystems on Earth* (Ed. Bock and Goode) Wiley and Sons Limited
Figs. 4.7 5.2	Courtesy of Lamont - Doherty Earth Observatory of Columbia University
Fig. 5.1	From the *Smithsonian Magazine*, July 2000 vol. 31 page 92
Fig. 5.3	Courtesy of Movile Cave Project, Romania
Fig. 5.4	Courtesy of National Oceanic & Atmospheric Administration
Fig. 5.5	Courtesy of National Science Foundation
Fig. 6.1	Redrawn from original figures by B U Haq in *Marine Geology and Oceanography of Arabian Sea and Coastal Pakistan* (Eds. Haq, BV; Milliman, JD) Van Nostrand Reinhold Company, New York
Fig. 6.6	Courtesy of Jorge Tutor.
Fig. 7.1	Courtesy of US Geological Survey
Fig. 7.2	Redrawn from original data from J J Sepowski jr
Fig. 7.3	By courtesy of 'coolrox'
Fig. 7.4	Courtesy of the Chicxulub Scientific Drilling Project core lab in the Museo de Geologia, Mexico City
Fig. 7.5	Courtesy of Geological Survey of Canada
Fig. 7.6	© Calvin J Hamilton, Views of the Solar System
Fig. 8.1	Courtesy of Starsem

Fig. 8.2   Courtesy of Beagle 2 project

Fig. 8.3   Courtesy of Lunar and Planetary Institute

Fig. 9.5   Courtesy of Alcatel Space Industries, Cannes

Fig. 9.6   Courtesy of Arecibo Observatory, National Astronomy and
                    Ionosphere Center, Puerto Rico

Fig. 9.7   Sagan and Drake (*Scientific American* vol. 232 page 80) and
                    reproduced with permission of Scientific American

Back cover   © 1993 Smithsonian Institution

Index